MW00442549

Military Interiors of the 19th Century

by

William L. Brown III

Cover design by Ryan C. Stouch

Printed in the United States of America

Published by THOMAS PUBLICATIONS
P.O. Box 3031
Gettysburg, PA 17325

ISBN-0-939631-42-3

CONTENTS

FOREWORD

Since the dawn of civilization, man has graphically recorded the world of soldiers. Images of victorious commanders and their armies engaged in the wars of empires have been common themes for artists illustrating great national events in bas-relief, paintings, prints and other similar mediums. Since these renderings were usually created to serve as visual narratives recounting important moments on the battlefield, they offer historians an extremely limited view of the total environment of a professional soldier. Exteriors and, most importantly, the interiors of military support structures erected to house, feed, supply, and heal were usually of little general interest and largely went undocumented for posterity until the third quarter of the nineteenth century. By this time, a new technology, photography, turned its eye upon the places where visual record had been largely confined to individual memory.

In 1839 the first widely practiced form of photography, the daguerreotype, was introduced in France, and Americans quickly embraced the new discovery. Daguerreotypes, unique images created upon silver plated copper sheets, were soon being made of citizens and soldiers within the confines of the studios of professional daguerreotypists. Due to the limitations of the technology which depended on natural light to illuminate subjects during the length of time (compared to today's fractions of seconds) needed to make an exposure and required a darkroom nearby to prepare the light sensitive plates, use of the medium outside of the studio was difficult. This was particularly true when attempting to record the interior of a room not specially constructed to allow the entrance of a sufficient amount of daylight. Consequently, America's military daguerrean record consists of studio and outdoor portraits of statically posed regulars and militia, and the process was employed during the Mexican War to record for the first time an American army in the field. But daguerreotypes of the interiors of the facilities used to support these and other soldiers of this era have yet to be discovered.

By the mid-1850s, a new photographic medium eclipsed the reign of the daguerreotype. The collodion wet-plate process used glass as the base for a light sensitive emulsion. Multiple paper prints (whose only color was applied by hand) could then be produced from a negative, which was an improvement over the daguerreotype's "one of a kind" limitations. Photographers, however, still had to depend on natural light for illumination and lengthy exposure times, though it was sometimes necessary to cover windows (Figures 3.14, 3.21, 4.4) in an interior view which would have overexposed portions of the negative. The light sensitive emulsion had to be exposed and developed while still tacky or "wet," again making the nearby darkroom a necessity. Cameramen used this cumbersome method to make their remarkable record of the Civil War as well as these postwar views of an officer's quarters (Figure 1.5) and enlisted men's barracks (Figure 3.10) reproduced here.

Due to the technical complexity and equipment cost, daguerrean and wet-plate photography was limited to the professional and serious amateur. Those that worked outside the studio's controlled environment were usually contracted by a specific customer, or the photographer thought that any resulting photographs would be of enough interest that they could be sold to the general public. Officers could better afford to use local galleries to record on-site their families and fashionably decorated living quarters than most enlisted men in their barracks. They had to wait for the government to take an interest in documenting their environment or for the advent of a more democratic photographic technology. Photographers wanting to reach a broader market sometimes used special twin lensed cameras that produced a double negative from which stereographs could be made. When observed through a special viewer, scenes on these double image cards appeared three-dimensional. Cameramen used stereography to record the Civil War, and some of their work (Figures 3.6, 4.6, 7.7) provide us with a rare glimpse of military interiors of that era.

By the 1880s, pre-manufactured "dry-plate" glass negatives were available that eliminated the technical and often precarious steps of the wet-plate process. Celluloid film and George Eastman's Kodak box camera soon followed. These developments opened photography to the amateur picture taker and ushered in the era of the commercially developed and printed snapshot. As film's light sensitivity increased along with the introduction of the electric and "flash" light technology, the ability to take indoor pictures led to an explosion in the number of military interior photographs after 1900.

Within these pages, William Brown has assembled the first comprehensive image collections of permanent and temporary interiors familiar to America's nineteenth century soldiers. Historians, curators, movie makers, military enthusiasts, and anyone interested in this dynamic period of our history should find this compilation a valuable aid when researching, interpreting, and understanding this often neglected chapter of American military history.

Ross J. Kelbaugh
Baltimore, MD

ACKNOWLEDGMENTS

Many people have been kind and helpful to me in my pursuit of photographs for this book. Among those to whom I owe the greatest thanks are Phil Porter, Mackinac Island State Park Commission; Myrna Williamson, State Historical Society of Wisconsin; Darrel Garwook, Kansas State Historical Society; Lynda S. Roper, Fort Sill Oklahoma; Mark McFarland, Library of the University of Oklahoma; Arthur L. Olivas, Museum of New Mexico; Les Jensen, Fort Sam Houston; John M. Manguso, Director, Fort Sam Houston; and Brian L. Dunningan, Director, Old Fort Niagara.

Other friends and colleagues who not only assisted me in finding and acquiring photographs, but also provided encouragement to push on include Rene Chartrand, Ottawa, Canada; Burt Kummerow, St. Mary's City, Maryland; Michael J. Winey and Randy Hachenberg at Carlisle Barracks, Pennsylvania; Doug McChristian and John Sutton at Fort Davis, Texas; Marilyn Wandrus, Springfield, Virginia; Bob Borrell of Washington D.C.; William Henry, St. Louis, Missouri; Paul. L. Hendren, Fort Union, Montana; Mike McAfee, West Point, New York.

Special thanks also go to Wayne Daniel, Fort Concho, Texas; Craig Nanos, Upper Darby, Pennsylvania; David Clary and Sarah Olson, and my colleagues in the Furnishing Division of the National Park Service at Harpers Ferry, West Virginia. I also send a very special thank you to Ms. Martha J. Straynoun, Editor for American Association for State and Local History. She was responsible for the initial editing and the side bars. I would also like to thank Candy Dunnigan who re-typed the entire manuscript onto a computer. And finally, to my wife Barbara and William and Robert for their encouragement and support. I also want to thank the Lord for each day.

INTRODUCTION

This book began, really, as a curiosity. I found a few photographs of the interior of some military barracks, seven or eight years ago, and I started looking for more, in a casual way. I've always been fascinated by nineteenth-century images and can remember many enjoyable hours spent with close friends and kindred spirits, pouring over images of street scenes that pictured such various details of bygone eras as clothing styles, advertising techniques, architecture, interior decorating, household furnishings, and a myriad of other features of daily living. What inspired me, along the way, was my job as a furnishings curator for the National Park Service. When one starts doing serious research on military furnishings, it becomes quite clear that very little information on that topic has been published.

I hope that this book will be helpful to those directly involved in refurnishing military structures characteristic of the nineteenth century. It has been my experience that viewing a large number of images of the same era and subject matter gives one a "feel" or a good *working impression* of that time period. Such a study of contemporary photographs of everyday objects common to an era can also be of great help to latter-day refurnishers in providing information about ways such objects were usually grouped together and placed within a room. Since such groupings and arrangements do change, markedly, over long periods of time, authentic contemporary photographs, such as the ones in this book, provide a direct and incomparably helpful link to the past that verbal descriptions, listings, and inventories of personal possessions could never do.

It has been my hope in putting this book together that those who examine these photographic keys to the past will not flip through the pages hurriedly, but will take time to look closely at the images that appear here, examining the objects shown and the rooms they appear, in order that the viewer, in a manner of speaking, may enter into these spaces.

One difficult aspect of working with old photographs is the need to use one's imagination to superimpose color on a black-and-white scene. Studies of the very complex subject of interrelationships between what we see and what we subconsciously believe indicate that the absence of color often causes the casual viewer of black-and-white photographs to form the notion that the period was totally drab. I've heard it said that many people's early impressions of Russia were of that sort, some viewers believing that everything Russian was all gray.

The need to imagine the colors used in furnishing the military quarters shown here may perhaps be less difficult than it might otherwise be, since, in most instances with these rooms, we do not need to speculate about what color the wallpaper should be—usually, there isn't any. Though wallpaper had been in use in Europe from the sixteenth century onward, and had been manufactured and used in Colonial America since the mid-eighteenth century, it was not often found in military quarters of the 1800s. One might theorize that the attempt to add color to real-life rooms, such as those on the pages that follow, help to account for the amazing variety of wallpaper now available.

I believe that anyone with the least bit of interest in material culture will find the time spent in examining these photographs to be time well spent, if for no other reason than to see and marvel at the amounts and kinds of objects that we Americans took west with us. Also—and this may be a most important point—one forms some idea of the commonality of the objects then in ordinary use. It is truly surprising to be reminded that most of the items we now consider essential to the convenience and comfort of daily living were nearly all invented, established or discovered in the nineteenth century. The information conveyed by the pictures in this book should be useful to both curators and historians—even those not necessarily involved solely with military research. I suspect that most readers with a real interest in nineteenth-century living will find the style and furnishings of our military officers' quarters of that era very similar to the style and furnishings of our nineteenth-century civilian middle-class homes.

FORT SMITH, ARKANSAS.
From *The Prairie Traveler.*

OFFICERS' QUARTERS

The details evident in many photographs in this book may cause some surprise to the reader who is familiar with the Hollywood version of the western army post of the nineteenth century. The men who resided in the military quarters pictured here, often accompanied by their wives and children, lived in remote, isolated, sparsely settled, and frequently hostile areas of the country, and it may be assumed that they lived with makeshift, temporary household furnishings and castoffs, rounded out by a few random hand-me-down items with which to make do. One might very easily assume that the only bits of the "civilized East" these military families would own would be a few treasured items small enough to be carried about with them in a single trunk. However, the photographs on the pages that follow reveal a totally different lifestyle.

On frontier duty assignments throughout the 1800s, American military officers and men tried to take with them as many of their own household furnishings and equipment as they could, as tangible reminders of the greater sophistication and more graceful living of the more settled areas they were leaving. Neither the men nor their wives had any desire to go native and rough it. Even in temporary quarters, they tried to keep up appearances.

Frances M. A. Roe, in her book *Army Letters from an Officer's Wife,* describes life in January 1873, at Cimarron Redoubt (in Indian Territory, present day Oklahoma), a fortification made of gunny sacks filled with sand. Mrs. Roe had the earthen floors of their quarters there leveled and

> the entire dirt floor covered with clean grain sacks that were held down smooth and tight by little pegs of wood and over this rough carpet we have three rugs we brought with us. At the small window are turkey-red curtains that make very good shades when let down at night. There are warm Army blankets on the camp bed and a folded red squaw blanket on the trunk. The stove is as bright and shining as the strong arm of a soldier could make it and on it is a little brass teakettle singing merrily.[1]

This home-making care was expended on a place where the Roes only spent three weeks in residence.

Most army officers of that era were entitled to the free personal service of an orderly, recruited from the enlisted men and often called a "striker," who served as butler and general houseman. Many officers hired additional household personnel to aid the officers' wives with daily chores. This was, one must remember, a time when such timesavers as vacuum cleaners, washing machines, electric irons, and dishwashers were things unheard of, and the endless work of keeping a home livable was done by hand.

Many officers, including Lieutenant Colonel George Armstrong Custer, felt that the daily grind of housekeeping was demeaning drudgery that officers' wives need not be subjected to. Libbie Custer wrote of the colonel that "Nothing seemed to annoy my husband more than to find me in a kitchen; he determinedly opposed it for years and begged me to make a promise that I would never go there for more than a moment.''[2]

Similarly, Alice Blackwood Baldwin, writing in 1867 and 1868, refers many times to ''our soldier cook.''[3]

One of Caroline Frey Winne's letters, written December 23, 1874, describes the Winnes' quarters at Sidney Barracks, Nebraska:

> We have a set of quarters assigned to us, not yet finished but they will be in a few days—a nice little house with hall and parlor and bedroom and dining room... There are two houses alike under one roof. There is a porch in front of the house. There is a back stairs and room over the kitchen for servants and over the front of the house two bedrooms with dormer windows. I will send you a little plan of the house when we get in it. We shall have the rooms measured and send for our carpets today.[4]

Happily for those who recognize and cherish the value of such records, Mrs. Winne did do a plan, in January 1875 (see Figure 1.1).

Officers' quarters at Fort Larned, Kansas, were described by Etta White, daughter of the post chaplain, David White, in a letter to her brother in January, 1878:

> The houses are all of stone one story high with verandas in front and at the side. A large hall runs through the center eight by thirty feet. On either side of the hall are large rooms fifteen feet square and double doors between. The wood work is all grained and varnished. The ceilings are the highest I ever saw, and the walls are all calcemined. Each room of a different color. We have ten rooms in all, and they all have some furniture in them. We have two movable wardrobes, three corner wardrobes, two fancy shelves, six or seven tables, cupboards, five in number. When we got here they had up three stoves and fire in them all and wood enough to do us for a month. There is only a part of a company here, only thirty-two men in all. The captain had them all working for us. There are only two officers in the fort beside Pa and the Dr. who is our next door neighbor. He called on us last night but I did not go in. Every man in the Fort is married, so you can imagine we are gay. There are two organs and two pianos in the Fort. There is a Library here of over a hundred volumes and we can get books out whenever we want to. I have not been there yet, and they take fifty dollars worth of dailies here for public reading. We had turkey for dinner. Don't you wish you had been here. Remember me to all who still remember who I be.[5]

John G. Bourke, a captain in the 3rd Cavalry, describes his quarters at Camp Grant, Arizona, in the 1880s:

> ...an apartment some fourteen by nine feet in area...nothing but a single cot, one rocking-chair. Visitors, when they came, generally sat on the side of the cot—a trunk, a shelf of books, a small pine wash-stand, over which hung a mirror of greenish hue, sold to me by the post trader with the assurance that it was French plate... There were two window-curtains, both of chintz; one concealed the dust and fly specks on the only window, and the other covered the row of pegs upon which hung sabre, forage cap, and uniform... There was one important article of furniture connected with the fireplace of which on occasion, I was wont to stir up the embers, and also to stir up the Mexican boy Esperidion, to whom, in the wilder freaks of my imagination, I was in the habit of alluding as my ''valet.''
>
> The quartermaster had recently received permission to expend ''a reasonable amount'' of paint upon the officers' quarters, provided that the same could be done ''by the labor of the troops.'' ...But the floor was rammed earth and not to be spoiled...a very acceptable present of a picture—one of Prang's framed chromos, a view of the Hudson near the mouth of Esopus Creek—which gave a luxurious finish to the whole business. Later on, after I had added an Apache bow and quiver, with its complement of arrows, one or two of the bright, cheery Navajo rugs, a row of bottles filled with select specimens of tarantulas, scorpions, rat-

tlesnakes, and others of fauna of the country, and hung upon the walls a suit of armor which had belonged to some Spanish foot-soldier of the sixteenth century, there was a sybaritic suggestiveness which made all that has been related to the splendors of Solomon and Sardanapalus commonplace...

But, be that as it may, the suit of armor—breast and back plates, gorget and helmet—nicely painted and varnished, and with every tiny brass button duly cleaned and polished with acid and ashes, added not a little to the looks of a den which without them would have been much more dismal.[6]

All in all, these officers and their families kept up the style of the period, and, in many instances, it would be difficult to tell that the dwellings pictured on the following pages were military quarters and not those of any middle-class civilian family in a settled area of the Victorian era. In many instances, the nearest railroads were hundreds of miles away. In viewing the rooms on the following pages, one should keep in mind the bleakness and isolation of the frontier country where these forts were established and the hostile environment in which the families who lived in them conducted their daily lives.

The kind of pluck that saw them through is expressed in these few sentences by an officer writing in The Army and Navy Journal, in 1868:

In a word, officers coming to the Plains should remember that they are going among gentlemen who live as such...as far as the circumstances of the case will admit...and not among a set of frontier ranchmen, who sleep on the ground and eat fried bacon from their fingers.[7]

It is not surprising that those officers and their wives made a great effort to transport civilization with them on their military assignments. They worked hard to keep up with current national news, with prevailing fashions, and particularly with manners and conversation. They probably worked harder at all aspects of "the civilized life" than did their urban cousins in the East. It must not be forgotten that, although they were on the frontier of civilization —and sometimes beyond that line—they were Victorians, and they thought, dressed, behaved, decorated their living quarters, and lived in a way that reflected the spirit of their times.

Concerning the photographs themselves—particularly those showing the officers' quarters—I am sure that many of these views were especially arranged for the camera. What is important, however, is not that the settings were arranged, but that the furnishings used in these arrangements were there, at that particular time and place—that their owners had brought them along and that they felt that this was the most tasteful way for them to be displayed. I suggest that these photographs, though they reflect the military life of those times, are a fair look at the prevailing trends guiding tastes in furnishings and fashions in civilian middle-class America of the nineteenth century as well, and that they are especially important since so many other Victorian interior views, both published and unpublished, show the homes of the upper classes only.

Figure 1.1. Officers' Quarters No. 4, Sidney Barracks, Fort McPherson, Nebraska, January 1875

An 1875 drawing by the wife of an officer at Fort McPherson, Nebraska, was the model for the modern-day drawing that appears in Figure 1.1. The adaptation was drawn to accompany an article called "Letters of Caroline Frey Winne from Sidney Barracks and Fort McPherson, Nebraska, 1874-1878," edited by Thomas R. Buecker and published in 1981 in the *Nebraska State Historical Society Magazine.* The original drawing was made by Caroline Winne, wife of Dr. Charles K. Winne assigned to Fort McPherson. Mr. Buecker says of the adaptation: "[This] drawing [was] adapted...from original by Caroline Winne. Officers' quarters Number 4, occupied by the Winnes, was the south half of a frame, adobe, brick-lined duplex built in 1871. Captain Hawley occupied quarters Number 3, to the north. Two upstairs bedrooms were probably not used by the Winnes. Caroline's notes on the rooms and furnishings accompanied the [original] drawing:

> (1) Bracket shelf with clock and pictures; (2) green-covered tables with pictures, books, and ink stand; (3) crimson-covered table with student's lamp and books; (4) fireplace mantel with pictures, vases, and two red stone Indian pipes; (5) dining room chairs, used throughout the house when not used at the table.

Pictures: Two dancing girls over table number 2; little slate pictures between two front windows; two of Dr. Winne's watercolor paintings on sides of windows. Gray carpet with crimson vine pattern in parlor and bedroom; striped matting in dining room. Crimson curtains at windows.

The living quarters of the Winnes are typical of those provided for officers at most permanent army posts of the late 1800s. Quarters like these survive today at Fort Larned, Kansas; Fort Laramie, Wyoming; and Fort Davis, Texas.

[Nebraska State Historical Society Magazine, 62, No. 1 (Spring, 1981)]

Figure 1.2. Officers' Quarters in a Casemate at Fortress Monroe, Virginia, ca. 1861-1865

A brick casemate at Fortress Monroe, Virginia, provided the homelike 1860s unit of officers' quarters seen in Figure 1.2. Casemates were secure fortress chambers, often vaulted, as this one is. They could be used as barracks rooms or as a battery for seacoast artillery. Fortress Monroe, a six-sided fort completely surrounded by a moat, was built between 1819 and 1834 and commanded the entrance to Chesapeake Bay and Hampton Roads. After the Civil War, Jefferson Davis, President of the Confederacy, was imprisoned in this casemate, from 1865 to 1867.

The carpet dominating the foreground, with its vivid figuration in Figure 1.2, appears to be a Brussels. A breakfast table dating from the 1840s serves as a parlor table. Here, it supports a parlor lamp (center) and two men's hats (at right), one for military dress, the other a civilian fedora. The lamp, a handsome one recognizable as an oil-burner, was most likely fueled by kerosene or coal oil, a thin oil that experimenters in the 1850s found could be obtained from either petroleum or shale oil. Kerosene was cheap, and after it was found to produce a clear, bright light when burned in lamps using a flat wick in an oil reservoir, it soon became plentiful. By the 1860s, it was being widely used to light homes, small shops, military buildings, and most structures in small towns and rural areas. Gaslight, used since the 1820s to light streets and public buildings in many U.S. cities, was too expensive to install and maintain in homes for any but the urban wealthy.[8]

Use of fabric to drape the mantelpiece in Figure 1.2 is a noteworthy touch, as are the precisely balanced mantel garnishments, a vase and a small decorative object at either end of the mantel. Feathers as well as fabrics were popular decorative devices of that era, as shown here on the left wall, in the background. The room seen indistinctly through the door at right might have been the officers' office; a furled and cased flag leans upright in the corner there, with a drum lying on the floor before it.

[Massachusetts Commandery, Military Order of the Loyal Legion, and the U.S. Army Military History Institute, Vol. 44, page 2172]

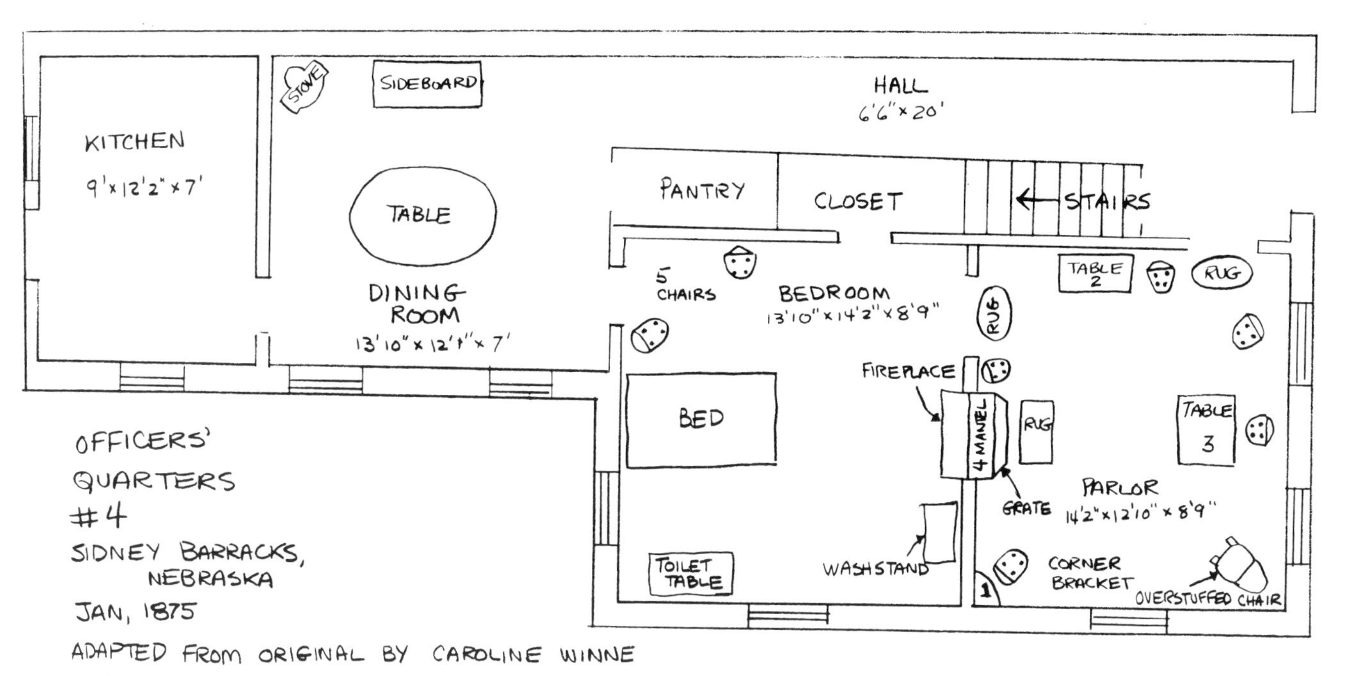

Figure 1.1

Figure 1.2

Figure 1.3

Figure 1.4

Figure 1.3. Lieutenant Granger Adam's Quarters, in a Casemate at Fortress Monroe, Virginia, ca. 1890

A quarter of a century later, here is the same room shown in Figure 1.2, very much brought up to the moment of the day it was photographed in 1890. In choice and arrangement of furnishings and objects displayed, the 1890s view of the room reflects a good bit of variety. The viewer's eye goes immediately to the comfortable-looking Morris chair at the left center; and one notices the portieres—heavy curtains hanging from rods over the doorways at left and right. Portieres had become quite popular by the last decade of the nineteenth century. They could be obtained in a wide variety of fabrics from mail-order houses all across the United States, and they were often quite decorative. They were also practical; they provided a bit of privacy in dwellings with large, multipurpose rooms that faced each other through big, open archways. In cold winter weather, they helped to conserve heat in what were almost always cold, drafty quarters.[9] Portieres were used in many military housing units of this era, and brackets for hanging them remain in many western forts. The decorative objects displayed on the mantel in this later view of the officers' casemate quarters at Fort Monroe are no longer arranged in the strict "balance" of those in Figure 1.2. An easel is used, here, to display selected works of art, one of them a classical engraving of Venus de Milo. The carpet appears to be a commercial Oriental type, probably made in England, although the manufacturing of carpets in America made a great spurt in growth after the power loom for weaving ingrain was introduced in the 1840s. By the end of that decade, American-made carpeting was readily available at fairly moderate cost. The imitation Oriental types of carpeting were very popular and were used extensively during the period of Colonial Revival from the 1870s throughout the early 1900s.

[Negative No. 1005, B.F. Browne Collection, National Archives, Washington, D.C.]

Figure 1.4. Officers Quarters at Fort Sumter, S.C. From a sketch by an officer on Major Anderson's staff. *Harper's Weekly,* Feb. 23, 1861

Two months after the publication of this sketch, this building, and presumably its contents, was destroyed and the country was at war. All West Point Cadets took courses in drawing, some of which appear in this book. Who the artist was is not clear, but he was a trained observer and almost every item in the sketch can be discerned. If this sketch is combined with the photographs in Figure 1.5, the reader gets a good view of the image of a bachelor officer's quarters at mid-century.

[The personal collection of the author]

Figure 1.5. Second Lieutenant Philip Reade, Company A, 3rd U.S. Infantry, Photographed in Quarters at Fort Dodge, Kansas, July 1867

Second Lieutenant Philip Reade, seated in a "Firehouse Windsor," resting his foot on what is probably a commode, looks toward the camera as he poses, in his room at Fort Dodge, Kansas (Figure 1.5). One says room, because Reade's modest rank—second lieutenants have always been low men on the army totem pole—entitled him to no more than a single room as his "quarters." In that room would have been his bed, dining table, parlor, and office. To the lieutenant's left is an officer's field desk, better than average in quality. In it he would have kept the records or papers of the company he commanded. This desk has a non-regulation, fancy, open-work gallery at the top and a fancy center section, as well. The front is shown opened, forming a writing surface. Topping the arrangement of military objects displayed on the wall above the desk is a cavalry sabre, partly out of its scabbard. Many infantry officers carried such sabres. Below this one are what appear to be two horsewhips and two more cavalry sabres. Two officers' dress-uniform sashes, an infantry bugle, and numerous Indian artifacts fill out the arrangement. An earthenware spittoon rests on the floor behind the lieutenant.

[National Park Service, Fort Larned, Kansas]

Figure 1.6. Quarters of Lieutenant Colonel George Armstrong Custer and Mrs. Custer, Fort Abraham Lincoln, Dakota Territory, 1873

The photographs in Figures 1.6 through 1.12 show the quarters of Lieutenant Colonel George Armstrong Custer and Mrs. Custer, in 1873, at Fort Abraham Lincoln, in the Dakota territory, where Colonel Custer was then stationed.

Like other wives of the nineteenth-century army men, Libbie Custer did her best to make the Custers' army post living quarters as relaxing and homelike as possible. That meant that, when assignments to new posts were made, the Custers—like their friends and acquaintances in the close-knit post-Civil War "army family"—packed up and took with them as many of their familiar treasured belongings as they could manage. They had plenty of practice in household moving; from the mid-1860s to the last decade of the century, army regulars assigned to frontier duty were spread thin and often shifted about, in the hundreds of miles of open, unsettled country along the western frontier. In *Boots and Saddles,* her book about those years, Libbie Custer says that "transportation for necessary household articles was often so limited it was sometimes a question whether anything that was not absolutely needed for the preservation of life should be taken with us."[10] The seven photographs of the Custers' Fort Lincoln quarters show clearly that Libbie Custer had the necessary inventiveness to make the best of it. Along with inventiveness, she must surely also have possessed the traditional gardener's "green thumb," to have coaxed her house plants at the fort into the luxuriant, leafy arrangement seen in the bay window (Figure 1.6) of the Custers' quarters. The arrangement is much like an idea mentioned in *American Woman's Home,* a book by Libbie's contemporaries, Catharine E. Beecher and Harriet Beecher Stowe,[11] and one wonders whether perhaps Libbie had the book in her library. Whether she had the book or not, Libbie had a firm knowledge of the basics of Victorian decorating. In addition to great numbers of potted plants, one also needed figured wallpaper, small pieces of statuary, souvenirs, and bric-a-brac (see Figures 1.8, 1.9, 1.11, 1.12). The guitar in the foreground in Figure 1.6 is a clue to the Custers' enjoyment of music and their familiarity with the instruments used to produce it.

[Custer National Battlefield Monument, Montana]

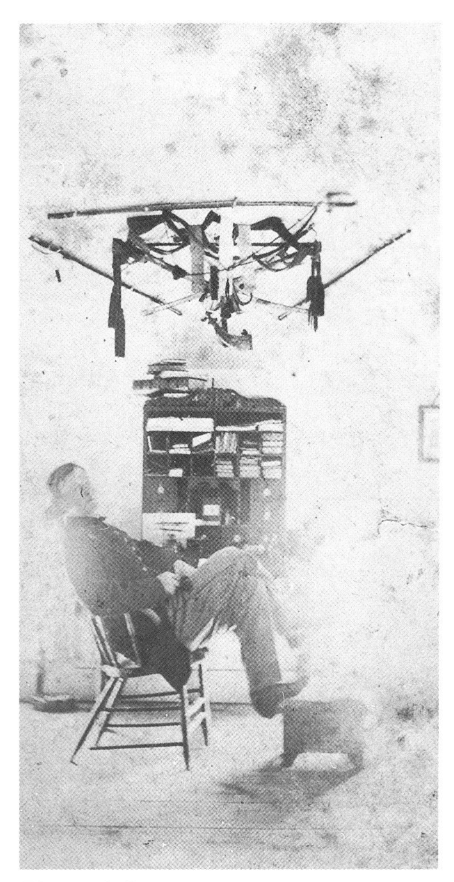

Figure 1.5

Figure 1.6

Figure 1.7

Figure 1.8

Figure 1.7. Quarters of Lieutenant Colonel George Armstrong Custer and Mrs. Custer, Fort Abraham Lincoln, Dakota Territory, 1873

Both Libbie Custer and the Colonel enjoyed music; and judging by the attentive group gathered about the handsome square grand piano and the fine harp in the Custers' parlor (Figure 1.7), they were enthusiastic participants in music-making as parlor entertainment for other army families and friends. In this view, the gentleman with the coveted post of musical-score-page-turner for the pianist looks very much like Colonel Custer himself.

The sharp-eyed viewer will notice, on the back wall of the parlor, the small, portable, two-plank book shelves suspended and held in place by sturdy ropes hung over a peg and attached to the shelves' four corners. This neat piece of equipment could be quickly and easily moved (see Figure 1.12).

[Custer National Battlefield Monument, Montana]

Figure 1.8. Quarters of Lieutenant Colonel George Armstrong Custer and Mrs. Custer, Fort Abraham Lincoln, Dakota Territory, 1873

A clearer view of the handsome, well-cared-for grand piano and harp belonging to Libbie and Colonel Custer appears in Figure 1.8. Here, the big musical instruments have been moved to a different area of the Custers' parlor. This view must have been made in the summer, since the piano appears to be standing in front of a screened—and most surely fireless—fireplace. At opposite ends of the mantel above the fireplace and the piano are two of the Custers' prized "Rogers Groups" —plaster table-top statuary pieces immensely popular throughout the country from the mid-1800s until after the turn of the century. A corner stand at left displays more of Libbie Custer's carefully tended green leafy house plants; and a handsome light fixture (upper right) hangs from the ceiling.

[Custer National Battlefield Monument, Montana]

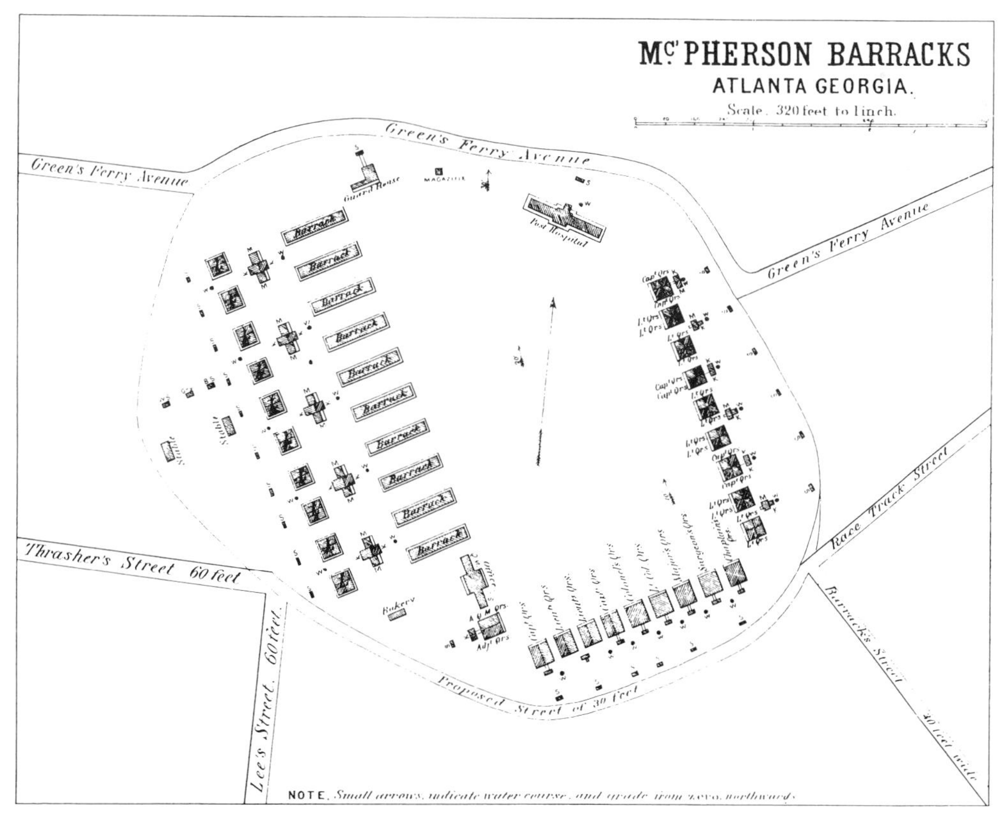

From Circular No. 4, *Report on Barracks and Hospitals.*

Figure 1.9. Quarters of Lieutenant Colonel George Armstrong Custer and Mrs. Custer, Fort Abraham Lincoln, Dakota Territory, 1873

Figures 1.9 through 1.12 show four areas of the library in the quarters of Colonel George Armstrong Custer and Mrs. Custer. The library housed many of the colonel's favorites among the family keepsakes—books, a huge map, family pictures, trophies of hunting— and the family's "Rogers Groups." In Figure 1.9, a pair of antlers mounted on a wooden plaque forms a decorative wall rack that supports the colonel's sword, field glasses, an everyday cap, a scarf, and a marvelously tall, lifelike stuffed bird that appears to be a sandhill crane. One of the "Rogers Groups" appears (center foreground) just below the wall rack. Another wall rack (at right) accommodates weapons—rifles, at upper and lower levels; pistols and a hunting knife on the rack's small shelf.

[Custer National Battlefield Monument, Montana]

Figure 1.10. Quarters of Lieutenant Colonel George Armstrong Custer and Mrs. Custer, Fort Abraham Lincoln, Dakota Territory, 1873

Colonel Custer is said to have been quite proud of the mounted grizzly bear's head trophy prominently featured (Figure 1.10, upper center) on the Custer's wall. The mounted bird just below the bear's head appears to be a handsomely marked hawk.

[Custer National Battlefield Monument, Montana]

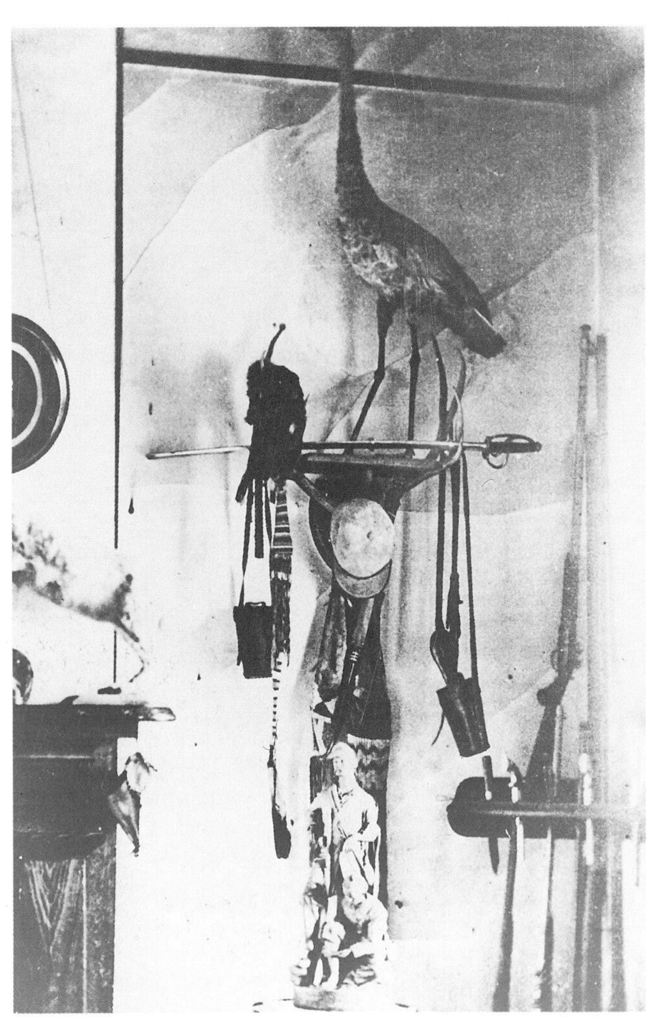

Figure 1.9

Figure 1.10

Figure 1.11. Quarters of Lieutenant Colonel George Armstrong Custer and Mrs. Custer, Fort Abraham Lincoln, Dakota Territory, 1873

Figure 1.11 provides a splendid close-up of Colonel Custer, reading at his library worktable. We also get a close look at a huge, framed map, shown here propped against the wall, but doubtless used as a wall hanging where space permitted. Two of the family's "Rogers Groups," seen earlier in Figures 1.8 (the taller figure) and 1.9, are also seen here; a handsome portrait of Colonel Custer decorates the wall above the map. Next to the portrait is a second "antler rack." This one, more elaborate than the smaller rack in Figure 1.9, provides the roost for a magnificent stuffed owl.

Both Figures 1.11 and 1.12 show wallpaper on the Custers' library walls, a decorative touch comparatively rare in military quarters of that era. Since the government did not provide wallpaper for military quarters in 1873, it must be assumed that Colonel Custer paid for this out of his own pocket.

[Custer National Battlefield Monument, Montana]

Figure 1.11

Figure 1.12. Quarters of Lieutenant Colonel George Armstrong Custer and Mrs. Custer, Fort Abraham Lincoln, Dakota Territory, 1873

Shot at a greater distance from the interior side wall of the library than Figure 1.11, Figure 1.12 shows the entire length of the library wall, revealing it to be a peculiarly angled structure with four "corners" visible—more angles than an ordinarily flat stretch of wall usually affords. We also note, with interest, the neat rope-hung book shelves at far left, looking very much like the shelves seen earlier on the parlor wall, in Figure 1.7. The smaller "antler rack" and gun rack, readily identified as having appeared earlier in Figure 1.9 on the unpapered wall of the parlor, also appear here. "Army ingenuity" devised many household furnishings that could be quickly and easily removed, packed, and transported to the next fort to which one was assigned.

[Custer National Battlefield Monument, Montana]

Figure 1.13. Second Lieutenant Robert J.C. Irvine, 11th U.S. Infantry, and Second Lieutenant Leighton Finley, 10th U.S. Cavalry, Fort Leavenworth, Kansas, Late 1870s

Figure 1.13 perfectly illustrates the quarters occupied by bachelor officers in the U.S. Army in the late 1870s. The viewer's eye goes immediately to the decorations on the back wall, particularly near the corners of the room, where a pair of elaborate wooden curios hang, holding an assortment of photographs. Photographs came to be used more and more in decorating military quarters as the turn of the century approached.

One notices that the walls in this room are not papered, and there is a fabric covering (foreground) on the floor. This floor covering does not appear to have any pattern and might very well be army blankets. I have heard the story of one officer's wife who sewed four gray army blankets together to make a carpet for her family's quarters. The lieutenants' parlor table (far right) has a good student lamp (a double one), an ink well, and numerous books. Additional lighting is available, as needed, from the kerosene lamp hanging, unlighted, from the ceiling (upper right).

The sideboard (center) is a curious item. Possibly made of shipping crates or trunks, it is covered with solid fabric that may perhaps be a blue wool. Along with the regular army uniform of blue kersey—a coarse, lightweight woolen cloth with a cotton warp—the army used some fabric of blue (often spoken of as "the national color") to cover government-issued tables and desks throughout the nineteenth century.

And one, of course, notices the numerous bottles on the floor, next to the water cooler, at right center. One reflects on the timeless brag of the "tough soldier" that drinking is his favorite pastime and assumes that these bottles held the lieutenants' favorite alcoholic potables. One also realizes, however, that while the occupants of these quarters may really have been living the uproarious life supposedly expected of the "carefree" young army officer, it is also quite possible that they carefully assembled that impressive bottle collection simply for show.

[Special Collections, University of Arizona Library, Tucson, Arizona]

Figure 1.12

Figure 1.13

Figure 1.14

Figure 1.15

Figures 1.14, 1.15, 1.16, 1.17, and 1.18. Quarters of Lieutenant Leighton Finley, 10th U.S. Cavalry, ca. 1880-1890

Figures 1.14 through 1.18 show a bachelor officer's quarters at different army posts in the late nineteenth century. The photographs were made by Lieutenant Leighton Finley, who lived in these rooms. Documentation is not clear about which quarters were at which posts. It appears that Figure 1.16 was at Fort Davis, Texas. It fits rooms surviving at that post, and Lieutenant Finley was stationed at Fort Davis in the early 1880s, along with the rest of the 10th Cavalry, one of two black cavalry regiments then in the U.S. Army.

Of major interest is the way in which Lieutenant Finley used the same furniture and decorative objects in various ways in different rooms during his career. We notice his wicker furniture; wicker was light, cheap, and easily transported, and was used by many military families of that era. We notice, also, that each time Lieutenant Finley was transferred to different quarters, he used the same decorative hanging parlor lamp; we see it suspended from the ceiling in all three of the parlor views (Figures 1.14, 1.15, and 1.16).

The records show that Lieutenant Finley never married, and thus the rooms he decorated and photographed show a bachelor's interpretations of the fashion of the period. The mantel decoration in Figures 1.14 and 1.18 could have come directly from a decorator's book of the period. It is doubtful that a furnishing curator would place pictures on the wall quite the way Lieutenant Finley did, without these kinds of views.

His use of fabric to cover, accent, and change the line of an object is interesting. There is little evidence here to let the viewer know that these rooms are the quarters of a military officer in a regiment seeing hard service against the Indians of the Southwest.

Figure 1.16

One notes that the book of etchings on the table covered with an Indian blanket in Figure 1.15 (far right) appears again in Figure 1.16 (center) on the same table, covered with the same Indian blanket. In Figure 1.15, the blanket has been tacked up a little way on the wall behind the book to form a sort of gallery. In Figure 1.16, the book shares table space with a bust that appears earlier on a corner shelf (far left) in Figure 1.14.

The dining room in Figures 1.17 and 1.18 appears, at first glance, to be the same room. A closer look, however, reveals that these views are of different rooms; the one in Figure 1.18 has a transom above the door and a fireplace mantel at far right. The portraits on the wall are Finley family pictures. In Figure 1.17, an Oriental paper umbrella is used as a decorative accent on the rear wall.

The sword in Figure 1.16 (center background) is a non-regulation officer's sword of the period 1840-1860. This one is the same sword shown in Plate 15 of Harold L. Peterson's book *The American Sword* (Philadelphia: Ray Riling Arms Books Company, 1954).

[Special Collections, University of Arizona Library, Tucson, Arizona]

WAR DEPARTMENT,

QUARTERMASTER GENERAL'S OFFICE.

Specifications for Barrack Bags.

Material.—To be made of six (6) ounce cotton duck, twenty-eight and one-half (28½) inches wide, dyed brown; to have drawing strings sixty (60) inches long, made of one-quarter (¼) inch braided cotton cord passing through two (2) sheet-brass grommets, one (1) on each side. To be sewed with No. 24 black cotton.

Size.—Thirty-two (32) inches deep and fifteen (15) inches in diameter.

Workmanship.—To be made flat stitched in the bottom and side seams; to have a tabling at the top two (2) inches wide for drawing strings, and conform to the sealed standard sample adopted this date.

Adopted March 9, 1885, *in lieu of those of March* 13, 1884, *which are hereby canceled.*

S. B. HOLABIRD,

Quartermaster General, U. S. A.

813—F., 1885.

Figure 1.17

Figure 1.18

Figure 1.19

Figure 1.20

Figures 1.19 and 1.20. Officers' Parlors, Fort Leavenworth, Kansas, ca. 1880

Parlors, or living rooms, typical of those seen in officers' quarters of the nineteenth century appear in Figures 1.19 and 1.20.

Brussels carpeting, quite popular in the the 1880s, covers the floor of the room in Figure 1.19, and a handsome set of portieres decorates the double doorway between the parlor and the smaller room adjoining it. No lamp appears on the parlor table; the double-lamp fixture suspended from the ceiling provides the necessary lighting. Ceiling fixtures such as this one were widely used during this period. The window hangings are distinctive, with the fabric at the top apparently chosen to complement the top section of the portieres at the doorway, and use of the folding paper fan as a decorative accessory above the picture on the wall at right center followed a trend of the era. The fans were considered symbolic of the imagined romance of "the mysterious East," and the new decor rage was items from the Japanese.

Figure 1.20 is a rare view of wallpaper in government quarters. While working on the furnishings at Fort Laramie (Wyoming), Fort Scott and Fort Larned (Kansas) and Fort Davis (Texas), no record could be found of any government purchases of wallpaper for officers' quarters. Since it was not government-issued, the wallpaper applied by the family who decorated this room would have been bought and paid for out of the family's own funds, and its appearance on their parlor walls indicates that they hoped to be in these quarters for some time.

Indian artifacts and military items are used decoratively on the wall to the right; and in this room also a double-lamp fixture—quite impressive—hangs from the ceiling. The parlor table lamp often seen in earlier years is missing.

[State Historical Society of Wisconsin. Photographs by Whitaker, WHI (X3) 9759, WHI (X3) 9758]

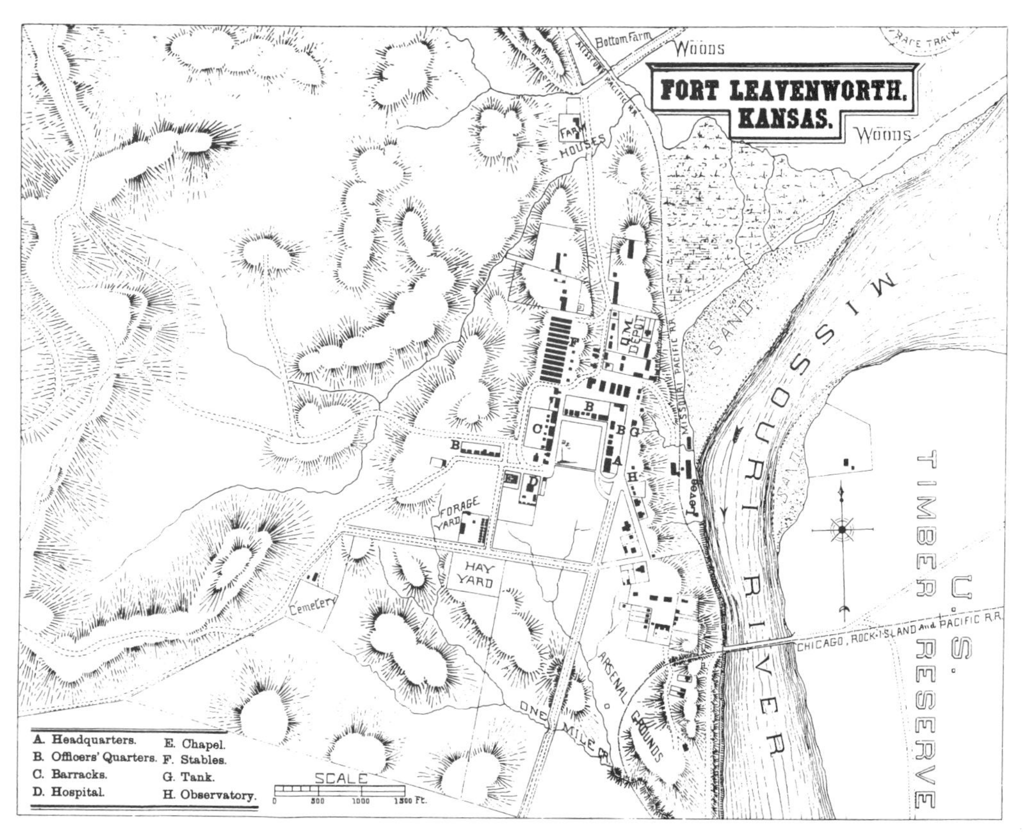

From Circular No. 8, *Report on Hygiene of the U.S. Army.*

Figures 1.21 and 1.22. Quarters of Second Lieutenant Lewis D. Greene, 7th U.S. Infantry, Camp Pilot Butte (formerly Rock Springs), Wyoming, 1885-1886 and April 1887

Figures 1.21 and 1.22 show the quarters of a junior officer, Lewis D. Greene. The open parlor doorway in Figure 1.21 affords a partial view of the bedroom with its ingrain carpet and a wall pocket hanging above the bedroom chair. A "wall pocket" was a Victorian wall ornament made of a backboard of polished wood—usually walnut—with a piece of needlework fastened to the front of it to form a pocket. All manner of things were tucked into these pockets, such as a current newspaper, a favorite picture, flat arrangements of dried flowers, small ferns, bits of brightly colored fabric, and all sorts of small treasures and keepsakes.

Two large, dark areas appear on the bedroom carpet. The nearer one is merely a blot of some sort on the photo print. The larger dark area, visible just at the foot of the bed, is the officer's pet dog at rest.

Greene's parlor served as an office as a well as living room. The parlor chairs (left and right) are folding chairs with rather ornate pierced backs; just visible in the far right corner is the arm of a wicker chair. The parlor lamp mat, probably put down originally as a simple cloth protector to keep lamp oil from seeping into the tabletop, has become a decorative divider. The numerous photographs in this parlor illustrate the growing popularity of photos as decorations for one's quarters.

Figure 1.22 shows Lieutenant Greene's dining room at the same post—the name of the post was changed from Rock Springs to Camp Pilot Butte on October 20, 1885. Three folding chairs with what look like caned backs appear in the dining room. Proudly displayed at right, on the serving stand, is the tea service. Greene's dining table probably also served as a work space. Officers and their families, assigned to living space according to the officer's rank, often had to make two or three rooms serve many purposes.

[The Library, U.S. Military Academy, West Point, New York]

WAR DEPARTMENT.

QUARTERMASTER GENERAL'S OFFICE.

Specifications for Barrack Chairs.

The legs, slats, and rungs to be of clear hickory, well seasoned. 1/4" or 5/16" iron on each side, front and back; countersunk head on one end and a nut on the other.

The seat to be of raw-hide, secured with raw-hide throng lacing beneath.

The chairs to be delivered "knocked down," *i. e.*, not put together, but packed in crates of convenient size for transportation by rail.

To be packed in boxes for shipment over the "Union Pacific Railway."

Adopted May 5, 1885, *in lieu of specifications adopted October* 22, 1883, *which are hereby canceled.*

S. B. HOLABIRD,
Quartermaster General, U. S. A.

1296—F., 1885.

Figure 1.21

Figure 1.22

Figure 1.23

Figure 1.24

Figure 1.23. Parlor of Lieutenant Edward B. Pratt and Family, Company K, 23rd U.S. Infantry, Stone Quarters, Fort Mackinac, Michigan, 1887

Lieutenant Pratt lived in this room as a young boy, when his father, Henry Clay commanded the post. His maternal grandfather, Captain John Clitz, also lived in this room in the 1830s when he commanded the Fort.

The multiple use of area carpets was very popular, as was the placement of furniture diagonally in the corners of the rooms. Naturally, the rocking chair, in this case wicker, was placed next to the parlor table, which had to have a small vase of flowers. The metal framed and upholstered relining chair was also a popular item. Note the use of fabric on the picture frames and mantel, also the Oriental shade on the back of the door.

The simple window treatment is also interesting. And a close look at the scene will reveal six-year-old Mary Lousie's dolly.

[Mackinac Island State Park Commission, Mackinac Island, Michigan]

Figures 1.24, 1.25, and 1.26. Quarters of Chaplain James A.M. La Tourrette, Fort Union, New Mexico, ca. 1890

Figures 1.24, 1.25 and 1.26 illustrate quite well the ways in which the nineteenth-century army officers and their families at remote frontier posts adopted that era's popular styles in home furnishings. We see here—again—the use of wicker furniture for the same reasons as noted earlier; wicker was inexpensive, light, and easy to transport. Wall hangings are a salient feature in both Figures 1.24 and 1.26. The introduction of hanging lamps allowed the center parlor table to be moved into a corner (Figure 1.24, right; Figure 1.26, left). Use of a fan as a decorative motif continued to be quite popular, and the mantel scarf, or lambrequin, in Figure 1.24, is of an unusually large size.

The photograph in Figure 1.25 appears to have been taken at an earlier date than its companion pieces in Figures 1.24 and 1.26. Figure 1.25 is a perfect example of the middle class parlor of the 1880s. The pump organ at left is only a step away from the center of all parlor activities, the parlor table, and several pieces of parlor furniture are early Eastlake, a furniture style that evolved in the 1870s, featuring rectangular, practical forms in oak or cherry, with geometric ornamentation and machined decoration of grooves and beveled edges. The carpet appears to be either Axminster or Brussels.

Figure 1.26 shows the difference that the passage of a few years' time made in the furnishings. The carpet in Figure 1.26 is different from the type shown in Figure 1.25; the table has been moved into a corner; and the decoration of the secretary and the colonial andirons in the fireplace foreshadow a growing interest in colonial revival as a result of the U.S. Centennial celebration. The Oriental influence is also strongly evident in this photograph.

[Negative Numbers 191, 192, and 193, Arrott Collection, Fort Union National Historic Park, New Mexico]

Figure 1.25

Figure 1.26

Figure 1.27. Quarters of Officers (unidentified), Fort Bowie, Arizona, ca. 1885

No carpeting is evident in the room shown in Figure 1.27, but we do see another example of a military officer's quarters with wallpaper. The bookshelves built over the mantel here are an unusual and distinctive feature, as is the window treatment, consisting of a curtain panel, drawn to the left side, as well as an outer drapery of some dark material with a flower print.

[Negative Number 24724-C, Arizona Historical Society Library, Tucson, Arizona]

Figure 1.27

Figures 1.28, 1.29, and 1.30. Quarters of Lieutenant Colonel Eugene B. Beaumont, Fort Bowie, Arizona, June 1886

Figure 1.28 is an exterior view of the handsome house that provided quarters for Colonel Eugene B. Beaumont, commanding officer of Fort Bowie, in 1886. Colonel Beaumont stands second from right in the group on the porch. This picture is included to show the kind of structure that held rooms like those shown in Figures 1.29, 1.30, and 1.31. Figures 1.29 and 1.30, photographed from different angles, show two views of the same room in the Beaumont house—the room directly behind Colonel Beaumont in Figure 1.28. Time brought about minor changes in the room's furnishings, evident in comparing the two shots. Both show the studio couch with fabric cover and pillows. Different scarves appear around the clock on the mantel in the two pictures; the dark scarf in Figure 1.29 is displaced in Figure 1.30 by a scarf of lighter color that immediately catches the viewer's eye. The corner shelf and the built-in bookcase with the protective drape are also positioned somewhat differently in the two shots. One notes the use of area rugs on the top of the carpet, more clearly seen in Figure 1.30. Comparison of the two shots makes evident the minor redecorating changes and the fussy little human touches that took place in this room which was lived in daily for some period of time.

[Negative Numbers 25617-8, 25606, and 25607, Arizona Historical Society, Tucson, Arizona]

Figure 1.28

Figure 1.29

Figure 1.30

Figure 1.31. Quarters of Lieutenant Colonel Eugene B. Beaumont, Fort Bowie, Arizona, 1886

Figure 1.31 shows a pleasant rear sitting room in the quarters of Colonel Eugene B. Beaumont in the Fort Bowie, Arizona, house shown in Figure 1.28. This room is much less formal, and the atmosphere is much more relaxed than that of the parlor shots in Figures 1.29 and 1.30. Here, again, area rugs appear about the carpeted area. The center table at right is a folding type, and may have been a camp table for field use. The pictures on the rear wall are hung from wires run from near the ceiling, as if the room had afforded picture molding. The pictures hang unobtrusively, with less of the forward slant from the top than pictures hung from wall pegs or nails. The cover on the circular table at left appears to be tacked down at the edge of the table top. The mustached man seated in the wicker chair at center is Colonel Beaumont. The officer to his right is still wearing his spurs, which he probably would not have done in the more formal front parlor of anyone's home, let alone that of the commanding officer.

[National Park Service Collection, Fort Bowie, Arizona]

Figure 1.31

Figures 1.32 and 1.33. Quarters of Captain Emmitt (Jack) Crawford, Fort Craig, New Mexico, ca. 1885

The Fort Craig, New Mexico, quarters of Captain Emmitt (Jack) Crawford appear in Figures 1.32 and 1.33. Captain Crawford was a hero to many who knew him, in both military and civilian life. Among his friends were Buffalo Bill Cody and other celebrated Western personalities of the late 1880s. Captain Crawford's quarters reflect a lively interest in current fashions and eclectic tastes of the period. The rooms were heated by a standard U.S. Army-issue stove, shown in both photographs of his parlor. A closer view of this same type of stove appears in Figure 1.24, in the quarters of Fort Union's Chaplain La Tourrette, also in New Mexico.

Captain Crawford's mirror (Figure 1.32, far right) shows the extreme angle at which some wall-hangings of that time were put up. Note the fabric knotted on the top of the balloon-backed chair in Figure 1.33. All the small treasures and personal touches evident here were used to add color and warmth to a drab environment.

Captain Crawford was killed in the line of duty, a year after these photographs were made. In pursuit of hostile forces in Mexico, he was wounded on January 11, 1886, and died of his wounds a week later.

[Negative Numbers 14512 and 14513, Museum of New Mexico, Santa Fe, New Mexico]

Figures 1.34 and 1.35. Quarters of Major Frank Baldwin, 5th U.S. Infantry, Fort Keogh, Montana, the 1890s

Seen in Major Frank Baldwin's quarters at Fort Keogh, Montana, in the late 1800s (Figure 1.34), are the double parlors common to many officers' quarters of that era. Many contemporary middle-class civilian homes also featured double parlors, or what were commonly called double parlors: two good sized rooms, adjoining with an open, double doorway, often arched, between them. Both rooms were not always used as parlors, except in the homes of the wealthy. One room nearly always did function as a formal "front parlor," but the other might be a less formal back parlor, a sitting room, library or bedroom, as needed.

Portieres, or door curtains, were very popular with people whose quarters—military or civilian—had double doors. (Portieres make their first appearance in this book in Figure 1.3.) These hanging panels of fabric—usually some heavy material, often velvet, almost always quite decorative—could be used to close off one double room from the other. This could be done either for a decorative display, for privacy, or, in cold weather, to help keep the room's warmth inside and the cold winter drafts out. In addition to the portieres, a portable screen on a wooden frame can also be seen in Figure 1.34. This screen seems to be propped against the open portiere panel at left, for use elsewhere as needed.

As in Captain Emmitt Crawford's quarters (Figures 1.32 and 1.33), the pictures on the wall in the Baldwin's quarters are hung at extreme angles. Much of the furniture has a protective fabric covering of some kind. The heavy rifle partly visible in the left foreground is not government-issue, but of a type that many officers preferred. The carpet, which did not photograph well, might be ingrain.

Figure 1.35 shows Major Baldwin's unusual collection of military artifacts, used here as a decorative wall display. European military objects are prominently featured, as are many artifacts of the Plains Indians. Fort Keogh, built in 1877, lay at the center of Plains Indian Territory. Keen-eyed viewers will notice a further display of swords and armor in the rear room of the Baldwin's quarters (far left) off the hallway.

[Custer Battlefield National Monument, Montana]

Figure 1.32

Figure 1.33

Figure 1.34

Figure 1.35

Figure 1.36

Figure 1.37

Figures 1.36 and 1.37. Quarters of Unidentified Officers, Fort Warren, Massachusetts, 1886

The Oriental influence popular in decorating during the 1880s is immediately evident in Figure 1.36; paper fans are spread decoratively across the transom of the doorway just off center, and a hanging scroll can be seen, through the doorway, in the next room. We see paper fans decorating the transom of the door in Figure 1.37, also, with an opened parasol hung near them on the wall for added emphasis. The better-than-average file desk seen through the doorway in Figure 1.36 is similar to the one shown earlier, in Figure 1.4.

Portieres appear at the doorways in both these photographs. In Figure 1.36, the portieres are pulled back, leaving the doorway open, with a single panel visible at left; in Figure 1.37, the portieres are closed, and we have a glimpse of the decorative edging at the top and side of the right panel. At the left (Figure 1.37), flanking the portieres, a quite decorative portable screen, on a wooden frame, has been spread open, quite possibly to show off its attractive covering. The frame is similar to that of the screen partially visible in Figure 1.34, though not quite so tall.

Fabrics are enthusiastically used as a decoration in these two photographs. In Figure 1.36, we see a swath draped over the frame of the picture at far right; below that, the student lamp appears to have fabric trim around the bottom of its milk-glass shade, with a decorative dark bow placed high on it in strong contrast. Figure 1.36 also reveals a straw matting on the floor of the room partially visible through the doorway. Lavish use of fabric for decoration continues in Figure 1.37, with the mantel scarf and the display shelves. The many artifacts exhibited there are also notable.

[U.S. Army Military History Institute, Carlisle, Pennsylvania]

WAR DEPARTMENT,
QUARTERMASTER GENERAL'S OFFICE.

Specifications for Bed Sacks.

Material.—To be made of cotton or linen drilling, or seven (7) ounce cotton duck of good quality.

Size.—Length, six (6) feet ten (10) inches; width, thirty-one and one-quarter (31¼) inches (measurements from corner to corner when filled); depth, four and one-half (4½) inches.

Opening.—To have an opening or fly in the center nineteen (19) inches in length, with one (1) by one and a quarter (1¼) inch stay-piece at each end; opening fastened with four (4) strings of three-quarter (¾) inch tape, placed equidistant from each end.

Finish.—All seams to be double; ends cut square; openings button-hole stitched at each end.

Adopted March 12, 1879.

M. C. MEIGS,
Quartermaster General,
Bvt. Major General, U. S. A.

337—O. M. G. O., 1879, Cl. and Eq. supply.

Figures 1.38, 1.39, 1.40, and 1.41. Quarters of First Lieutenant John Theodore French, Jr., 4th U.S. Artillery, Fort Preble, Maine, 1886

Figures 1.38, 1.39, 1.40, and 1.41 present a panoramic view of the parlor and dining room of the quarters of Lieutenant John Theodore French and his family at Fort Preble, Maine, in 1886. One reacts immediately to the richness of decorative detail in these rooms.

The walls, unpapered, are almost entirely covered with a wealth of personal touches and cherished possessions on display. There are pictures of all sizes, some elaborately framed. Many decorate the wall space above the piano (Figures 1.39 and 1.40). Pictures adorn a parlor shelf and side tables as well (Figures 1.38 and 1.40).

A hint of the Orient appears in all four of these photographs, in the folding fan on the wall above the piano (Figure 1.40), the opened parasol above the roll-top desk in the dining room (Figure 1.41), and in glimpses of the imitation Oriental carpeting on the parlor floor (Figures 1.38 and 1.39). These pseudo-Oriental coverings, with their distinctive abstract geometrical designs in bright colors, were very popular among the Victorians. They were often called "Turkey carpets," because of the many genuine Oriental carpets made in Asian Turkey.

In keeping with the trend of that day, swatches of fabric and small decorative hangings made of textiles are used ornamentally here for color. A short length of fringed material with a printed design drapes forward at one end of the piano, which has a full-length cloth runner on its top (Figures 1.39 and 1.40). Other lengths of colorful cloth hang from one end of the parlor shelf (Figure 1.38). A cloth handbag with a flowered design hangs from the top of the parlor bookcase (Figure 1.40). Another swings from a doorknob (Figure 1.38), and a long fabric bag is suspended from a decorative wall ornament between the portiere and the piano in Figure 1.39. Several lengths of fabric bearing colorful designs can be seen in all four photographs, mounted on dowels and hung on parlor walls or on the back of a door. Fabric trim appears on the shades of the lamps at the piano and on the round and oblong tables in the parlor (Figures 1.38, 1.39 and 1.40). The parlor window has lace curtains and, to increase the sense of wall space as well as to close off one area from another, portieres hang at the double doors between parlor and dining room, and between the dining room and kitchen. There appear to be "pull shades" behind the curtains at the parlor window.

As in many officers' quarters of the Victorian period, wicker furniture is used—here, in the company of a Huntzinger chair—the folding chair with fringed fabric in the seat at far right in Figure 1.39. These chairs—light and generally collapsible, though some of them did not fold—were named for the maker, George Huntzinger of New York City. Beginning in the 1860s, Huntzinger's company produced a wide variety of innovative furniture, and the "Huntzinger chairs" remained popular until the 1890s.

The piano organ that ranked as a kind of status symbol in the proper Victorian parlor was an even more important possession to military families in isolated army posts. In those days, many women of some social attainment could play the piano or the organ fairly well—it was part of their upbringing, like good posture and "pretty manners." Officers' wives, who possessed those skills and other social graces expected of those who ranked a bit above the crowd on the community's social scale, could be of great help to their husbands' career hopes. [French Collection, U.S. Army Military History Institute, Carlisle, Pennsylvania]

Figure 1.38

Figure 1.39

Figure 1.40

Figure 1.41

Figure 1.42. Frontier Tea Party, 1889-1893

In Figure 1.42, the photographer captured a marvelously unposed-looking shot of the officers' ladies at tea at an unidentified frontier post of the late 1800s. Social life at frontier garrisons was what the military community could make of it, and the officers' wives worked very hard at not losing touch with niceties. Whatever the occasion for the tea party, it called for snowy napery, a live floral centerpiece, floral decorations for the hanging lamp, and the rod above the doorway where the portiere usually hung. It required the hostess's silver tea service—and probably that of her best friend as well—and best party dresses for all the ladies. The butler for the festivities, seen at left offering delicacies to guests at the end of the the table, is probably the family's "striker," the enlisted man detailed to work as houseman for an officer's household. Each officer was assigned one striker for such duties.

[Charles D. Rhodes Collection, The Library U.S. Military Academy, West Point, New York]

Figure 1.42

Figure 1.43. Unidentified Officer at a Post in New Mexico, 1890

The photographic image is blurred, and no identification has been found for the man in Figure 1.43, seated, in civilian clothes, at a crude table that takes up much of the room's space. We do know that, although he is wearing civilian clothes, he was an army officer, and the photograph was made in his quarters. The room shown probably served as both parlor and dining room. Note the sideboard in the corner (center) and the folding lounge chair (right foreground), covered with a piece of cloth. The pictures on the wall appear to be tacked there.

[Negative Number FV-56-14-106-H-13-2, Fort Verde State Historic Park, Camp Verde, Arizona]

Figure 1.44. Parlor, Officers' Quarters, Unidentified Post in New Mexico, 1887

A parlor typical of those found in army officers' quarters in the 1800s appears in Figure 1.44. Salient points are the parlor table, with its bulky fabric covering, and the enormous lamp that had been placed on it. The portieres here are hung differently from the usual method; they look almost as if they were doors. The doorway that they are to screen opens into a hall. Visible on the table, at far left, is a hat, probably that of the gentleman present in the room.

[Negative Number FV-57-10-2-H-17-16, Fort Verde State Historic Park, Camp Verde, Arizona]

Figure 1.43

Figure 1.44

Figure 1.45

Figure 1.46

Figure 1.45. Kitchen, Officer's Quarters, Fort Verde, Arizona, 1890

Figure 1.45 is the only photograph in this book that shows the kitchen of a nineteenth-century U.S. Army officer's quarters in everyday use. Identification of the aproned woman at the work table is not available, however, she was probably a servant. Both formal and informal records and correspondence of officers and their families indicate that most officers preferred that their wives stay out of the kitchen. It may have been a point of pride on the part of the man of the house, or it may have been a matter of status—at any rate, it was a preference that could readily be indulged. This was the era in which help was cheap, and either enlisted men, their wives, or servant girls were nearly always available for cooking, laundering and cleaning in the officers' households.

The work table in the kitchen shown here was probably covered with oilcloth. All cooking for this household was done on the huge mid-century iron cookstove at left. Just to the right of the stove's handsomely decorated oven door is the firebox, where the fire was fed with stove wood sticks to keep it burning, winter and summer. Above the work table, a highly utilitarian clothesline stretches from wall to wall, a convenient height for dish cloths and towels.

The room seen through the open doorway at far right might have been the serving woman's bedroom, if she was a live-in maid, which was often the case in those days. Many officers' wives also brought hired young women from home with them on frontier assignments. Many of these girls later married enlisted men whom they met at the post.

[Negative Number FV-56-14-55-H-13-4, Fort Verde State Historic Park, Camp Verde, Arizona]

Figure 1.46. Officers' Quarters, Fort Union, New Mexico, 1880s

One sees striking similarities in comparing the officers' quarters in Figure 1.46 to the quarters of George and Libbie Custer shown earlier in Figures 1.6-1.12. Immediately noticeable similarities are the piano and the harp, the family portraits hung here from the picture molding on the rear wall, the use of wallpaper, the corner whatnot filled with bric-a-brac (far left), the carpet in the foreground, and the lengths of fabric covering the piano and the parlor table. Of similar comparison are the parlor table in Figure 1.46 to the one shown next, in Figure 1.47.

[Negative Number FV-56-14-27-11-17-4, Fort Verde State Historic Park, Camp Verde, Arizona]

Figure 1.47. Parlor, Officers' Quarters, Fort Verde, Arizona, 1890

A neat, plain parlor appears in Figure 1.47, in what are probably the quarters of a junior officer. The wicker chairs and the ingrain carpet are here, plus an area rug. The piano in the background at far right was a most important possession. A piano in one's house was, of course, partly a status symbol, but it was also a familiar and very popular means of entertainment and a real drawing card for social gatherings. At far left is the parlor stove. No curtains or draperies appear at the window, and the shutters are closed. Judging by the way the people pictured here are dressed, the season is probably late spring or early summer. The arrangement of the cloth on the parlor table is interesting, as it does not completely drape over all four sides, which was most common.

[Negative Number FV-56-14-105-H-13-3, Fort Verde State Historic Park, Camp Verde, Arizona]

Figure 1.48. First Lieutenant Charles R. Noyes, 9th U.S. Infantry, at Home, Madison Barracks, New York, 1892-1898

The parlor in the quarters of Lieutenant Charles R. Noyes and family, at Madison Barracks, New York, when this picture was taken in the 1890s, projected a quite refreshingly Spartan look. The dining room chair at far left is being used here in the parlor as a side chair, along with two rockers—one of them made of popular wicker. The single floral decoration on the center table is remarkably effective, and, in the absence of a table cover, the handsome table itself is decorative. Area rugs scattered about provide a splash of color. The window treatment, using blinds and lace curtains, and the carefully limited selection of pictures on the walls are at one with the clean, uncluttered, "controlled" look of this pleasant room.

[Noyes Papers, Special Collections Division, U.S. Military Academy, West Point, New York]

WAR DEPARTMENT,
QUARTERMASTER GENERAL'S OFFICE.

Specifications for Card Receivers.

To be made of tin, and in form according to the standard sample. The upper front and edges to be japanned emerald green. Height at center about three and five-eighths (3⅝) inches, width about three and three-fourths (3¾) inches. The edge at sides and bottom to be turned over front about one-eighth (⅛) of an inch full, leaving sufficient room to allow the sliding of an ordinary bristol-board card. Height of turned-over edges at sides about two and three-fourths (2¾) inches. A round hole three-sixteenths (3/16) of an inch in diameter in the upper edge at center.

To be fully equal to the standard sample in quality and finish.

Adopted February 20, 1882.

D. H. RUCKER,
Quartermaster General, U. S. A.

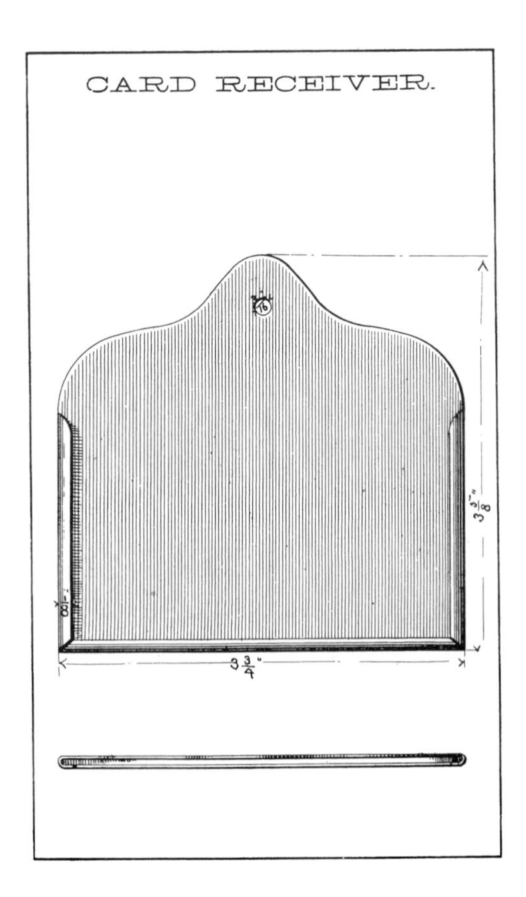

Figure 1.47

Figure 1.48

Figure 1.49

Figure 1.50

Figure 1.49. Mrs. Charles R. Noyes in the Noyes' Quarters at Madison Barracks, New York, 1892-1898

Gertrude (Mrs. Charles R.) Noyes faces the camera from the handsome desk that was also one of the furnishings of the parlor shown in figure 1.48. In this view, we get a closer look at what must be another window in the Noyes' parlor, this one with shutters instead of blinds, and with a potted plant flourishing on the window sill. A distinctive feature of this view—and one indicative of this family's absorption of the literary classics—is the single ornament atop the desk: a small, elegant statue of the mythological Roman deity, Mercury, messenger of the gods, with his winged helmet and winged sandals, bearing the caduceus with which he is often portrayed.

[Noyes Papers, Special Collections Division, U.S. Military Academy Library, West Point, New York]

Figure 1.50. Quarters of Second Lieutenant Benjamin C. Morse, 23rd U.S. Infantry, Stone Quarters, Fort Mackinac, Michigan, 1890

The parlor in Figure 1.50 was that of Lieutenant Benjamin Morse and his wife, then newlyweds. The room seems large, considering the amount of furniture we see in it, and the furniture seems to have been carefully pushed back out of the way, as if to show off the nicely patterned ingrain carpet with area rugs at left foreground and right background. A cotton spread in a print that resembles the pattern of the carpet covers the daybed at left. One notices the two folding chairs (far right), the placing and "outhang angle" of the pictures on the walls, and the draw curtains on the bookcase (foreground, far right). Shirred curtains such as these were often hung on bookcases or book shelves, to protect the book bindings against damage from exposure to light.

[Mackinac Island State Park Commission, Mackinac Island, Michigan]

WAR DEPARTMENT.

QUARTERMASTER GENERAL'S OFFICE.

Specifications for Mattresses.

To be equal in all respects to the sealed standard sample.

Material.—To be made of narrow stripe blue and white ticking; "herring-bone" or "twill" weave; the filling to be of good cotton linters.

Dimensions.—To be six (6) feet six (6) inches long, and two (2) feet eight (8) inches wide, and three and one-half (3½) inches deep when made up, and to weigh not less than twenty (20) pounds.

Workmanship.—To be made in a neat and substantial manner; to have twenty-eight (28) tufts of suitable quality of leather on the top and bottom.

Adopted March 28, 1885.

S. B. HOLABIRD,
Quartermaster General, U. S. A.

1075—F., 1885.

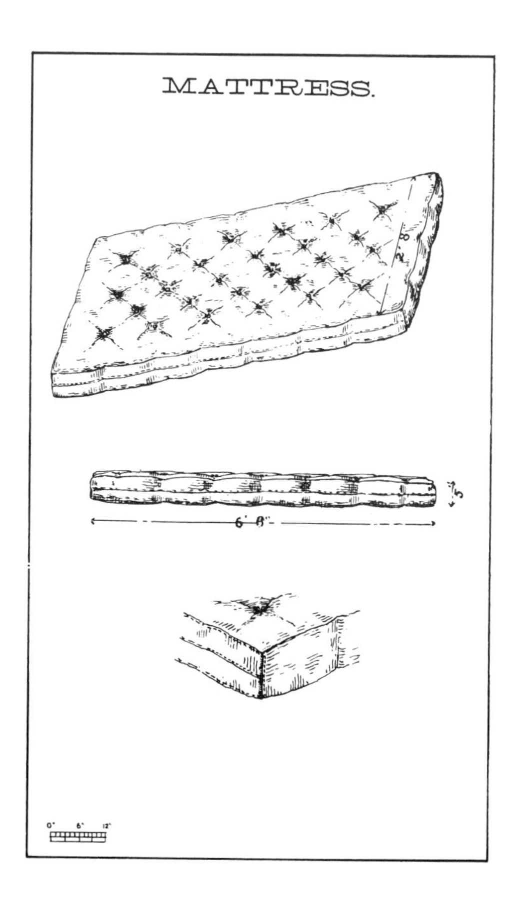

Figures 1.51, 1.52, and 1.53. Quarters of Lieutenant Hamilton, Fort Robinson, Nebraska, 1895-1897

The trio of photographs in Figures 1.51, 1.52 and 1.53 show not only highly individualistic views of an 1890s parlor and dining room, they show the photographer who made them as well. The rooms were in the quarters of a Lieutenant Hamilton, stationed from 1895 to 1897 at Fort Robinson, Nebraska. The photographer (see Figure 1.52) was Kate Hamilton, the lieutenant's wife. Mrs. Hamilton not only operated the camera that made the negatives for all three photos, she also developed the negatives and made the prints that we see—not ordinary accomplishments for a young officer's wife, in those days. Her pictures, including the self-portrait in Figure 1.52, lead one to conclude that Mrs. Hamilton was no ordinary young woman, but one with unusual fondness for books and the information they contained, one who enthusiastically—and creatively—kept up with the times. Decorating these rooms at the turn of the century, she used many ideas which were just then coming into popularity. After almost fifty years of Victorian clutter, people in the 1890s were moving into a home-decorating trend that was both more Spartan and less formal; more open space and greater tidiness, a sense of selectivity and less of a jumbled, "crammed-full" appearance, and more comfortable, welcoming touches intended to make guests feel at ease. Instead of a great display of flourishing house plants in this parlor, there is a small elegant arrangement of fresh flowers (Figure 1.51). The family's musical instruments appear to have been just casually laid down and remain close at hand; soft, decorative lounging pillows welcome the visitor; an elegant tea set is close by, and there is lavish use of colorful fabrics. Although Kate Hamilton's pictures appear in black and white, the viewer readily forms mental pictures of the splashes of color that must have appeared in the printed fabrics so skillfully sewn and displayed in the Hamilton parlor and dining room. A shirred fabric makes an inventive backing for the objects displayed in the corner shelf in Figure 1.51; and the sprightly print lining the corner china cupboard in Figure 1.53 seems to have been specially selected to enhance the china pattern. The cupboard was probably a built-in storage arrangement, as were many such units.

The Hamiltons' handsome china tea set appears, in use, twice—in Figure 1.51 at far left, and again in Figure 1.52, two years later. Figure 1.53 lends the impression that more pictures have been hung on the walls since the first parlor shot was made; and in Figure 1.53, one glimpses the polished wood of the floor, with a tiny bit of the end of an area rug in evidence, but no carpeting.

[Numbers H216-17, H216.5-10 and H216.5-1, Nebraska State Historical Society, Omaha, Nebraska]

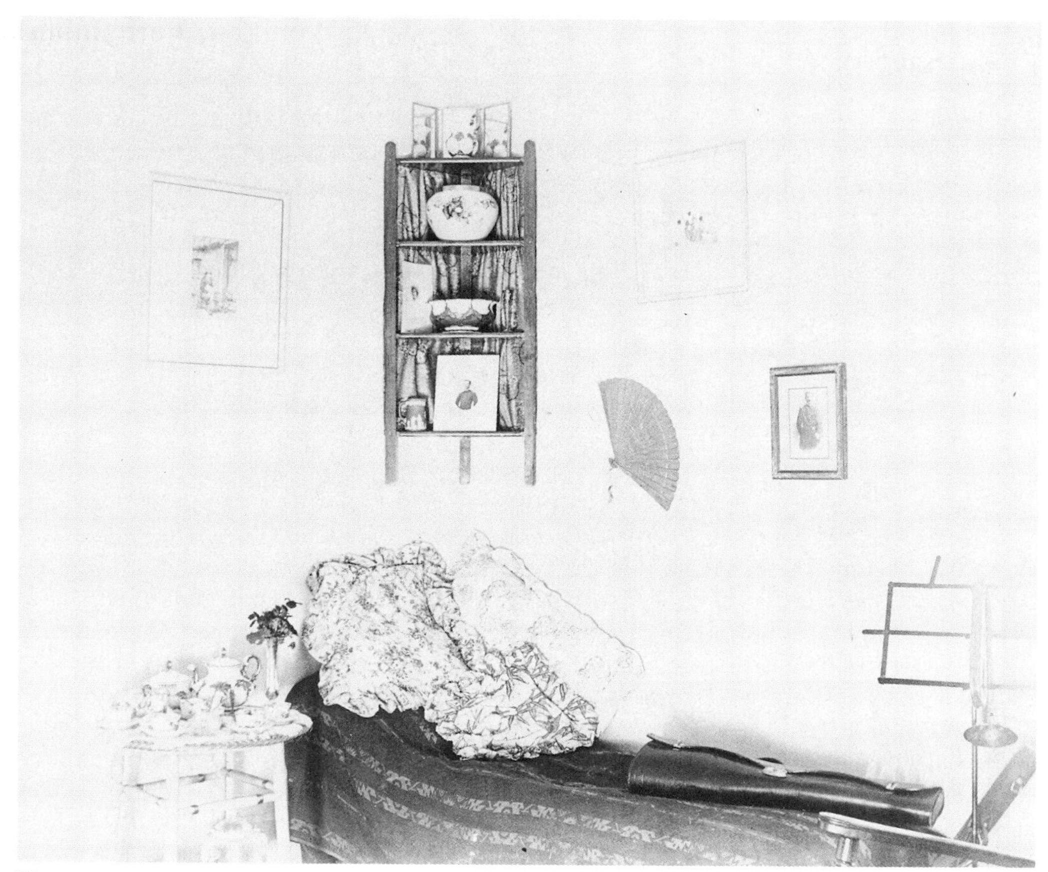

Figure 1.51

Figure 1.52

Figure 1.53

Figure 1.54. Dining Room, Captain's Quarters, Fort Sheridan, Illinois, 1893-1895

Unless one were specifically and thoroughly trained in the history of military housing, one would never guess that the pleasant, spacious dining room in Figure 1.54 is what it is: the dining room in the military quarters of a U.S. Army captain stationed at Fort Sheridan, Illinois, from 1893 to 1895. The captain, whose name does not appear among the data for this photograph, is seated at far left. If he had been dressed in civilian clothes instead of in uniform when the photograph was made, it would have been impossible to tell that this attractive room, with its handsome hanging lamp and ample, glassed-in shelves, was not a middle-class civilian dining room anywhere in the United States at the turn of the century. Actually, the captain and his family were lucky in having been assigned to Fort Sheridan at that particular time. Fort Sheridan was then one of the newer army posts, having been constructed and opened for service in 1887.

[Negative Number 92-F-61B-18, National Archives, Washington, D.C.]

Figure 1.54

Figures 1.55, 1.56, 1.57, and 1.58. Quarters of Chaplain Brant C. Hammond, Fort Sill, Oklahoma, Territory, ca. 1900

The posed pictures in Figures 1.55, 1.56, 1.57, and 1.58, taken at the turn of the century, show family members gathered in the parlor of Chaplain Brant C. Hammond, then stationed at Fort Sill, Oklahoma Territory. One of the most interesting things about these four photographs is their emphasis on family participation in various types of parlor entertainment enjoyed by the Victorian household: books, conversation, parlor table games, and the always-popular music.

Evident in the Hammond home is a strong interest in reading: well-stocked wooden bookcases appear on both sides of the parlor fireplace (Figures 1.56 and 1.57). The viewer deduces that reading was stressed, not only for individual enjoyment—as shown by Mrs. Hammond, the chaplain's wife (Figure 1.56)—but also for its value as a learning and research tool for younger members of the family, illustrated by Grace (left) and Blanche Hammond, the chaplain's daughters, in Figure 1.57. The number of books that military families took with them to frontier army posts in the 1800s may seem amazing to those of us accustomed to today's instant communications media, but to the army officers and their families of that earlier time, books and the many newspapers and magazines to which they subscribed were their cherished and only keys to what was going on in the world, and they were highly valued. As a group, these people were far more well-read, generally, than today's average American, and they kept themselves quite well-informed about national affairs and current trends and fashions. The art of conversation then was a valued social attainment and a far more lively pastime than it is today. The Victorians kept it going through lively exchanges of views with friends, often involving ideas discovered in reading.

In Figure 1.55, the Hammonds pause at the game table long enough to have their picture made: from left, Blanche, the chaplain, Grace, and the girls' younger brother, busily concentrating on game strategy though still in knee breeches.

The family's interest in music is clearly shown in Figure 1.58, as the Hammonds' youngest daughter, Ruth, seated at the organ, selects sheet music from the much-used music stand at right.

Decorative items seen about the Hammond quarters include handsome examples of Indian basketry, pottery, and fabric design (Figures 1.56 and 1.57). Indian artifacts were often used very decoratively in the quarters of most army officers of that era and reflected the art of the Indians living in areas where officers had been stationed. Nearly all the officers of the peacetime army had at least one assignment of frontier duty in Indian Territory.

[Negative Numbers 58-20-2, 58-20-24, 58-20-15, and 58-20-7, Fort Sill Museum, Fort Sill, Oklahoma]

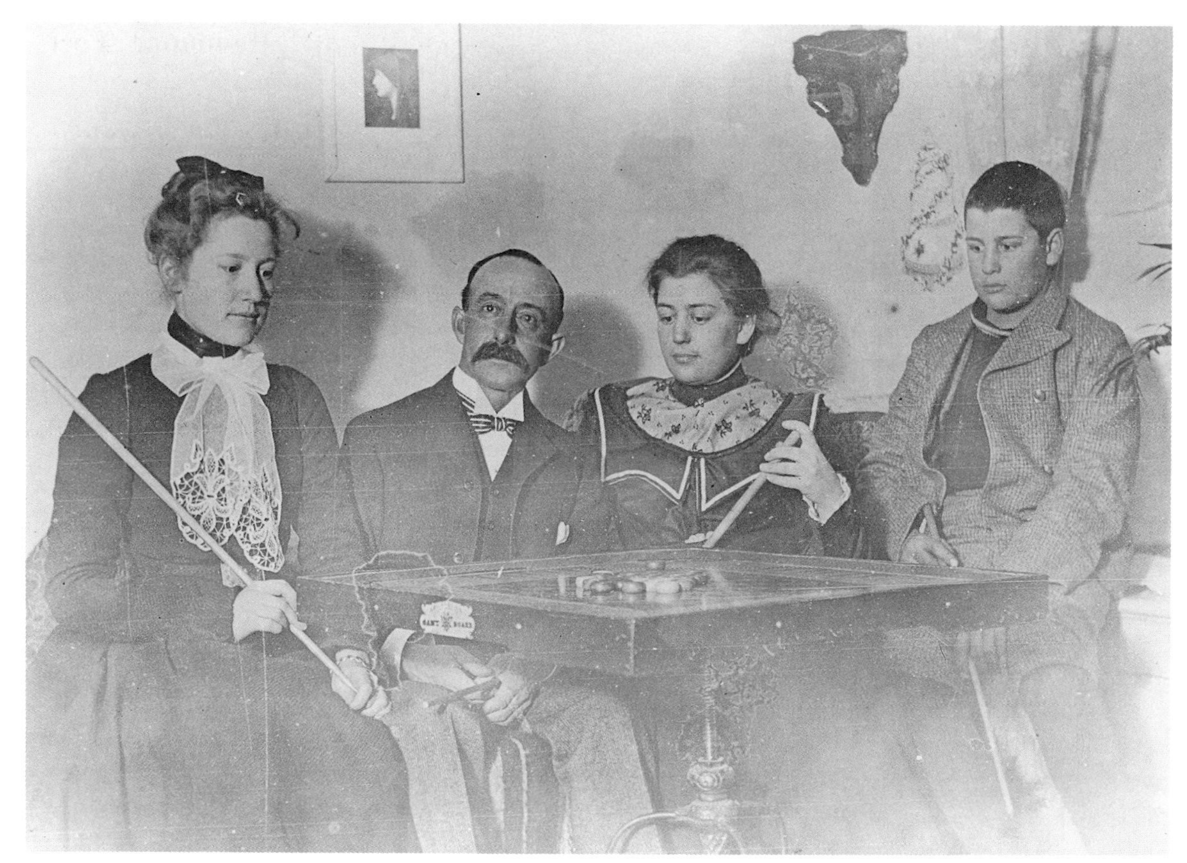

Figure 1.55

Figure 1.56

Figure 1.57

Figure 1.58

Figures 1.59 and 1.60. Parlor and Dining Room of Captain W. McCorliss, Gloucester, MA, ca. 1885

A classic middle class parlor of the late nineteenth century. The position of the sofa across the corner was a very popular way to arrange a long piece of furniture. The sofa and two chairs are part of a prolor set and are upholstered in horse hair. The arm chair appears to be an early lounge chair, and the adjustment knob is clearly in view near the right arm. The marble-top center table is no longer covered and does need a parlor lamp since kerosene now does the job better. The table contains a bowl of fruit and the family photograph album. Typically, the box-style grand piano dominates the room. The large engravings are all hung from picture molding and hang out from the wall. The carpet is probably ingrain, although a heart rug is used under the piano stool.

The use of ceramic objects on the shelf of the rear in symmetrical arrangement was recommended at this period.

Figure 1.59

Due to lack of space, the dining room also doubled as the morning room, or family room. The leaves have been removed from the dining table and a kerosene lamp is being used for reading and studying. The wallpaper, rare in military interiors, is probably of a maroon background since reds were the recommended shades for dining rooms.

The use of plants around the windows was another very popular custom. (Note Libbie Custer's rooms in Figure 1.7.) The use of area rugs scattered on top of the carpets was another common decorative arrangement.

Picture spikes with porcelain heads are used to hang the pictures in this room. The horsehair-covered lounging couch can be seen also from the parlor and appears to be a piece made during the mid-century. These two rooms appear to be almost identical in size and construction to a surviving officer's quarters at Fort Laramie, Wyoming. They were probably constructed according to standard army plans.

[Negative Numbers 22 CF-329 and 22 CF-330, National Archives, Washington, D.C.]

Figure 1.60

QUARTERS OF MARRIED AND SENIOR ENLISTED MEN

Figures 2.1, 2.2, 2.3, and 2.4 illustrate the level of military housing just below that for officers in the army command structure, quarters to which noncommissioned officers—sergeants and corporals— and married enlisted men, were assigned.

There was a great contrast between the housing and the life styles of these men and those of the officers, particularly for the wives of the enlisted personnel. Life was hard for the common soldier's wife, and the lower the soldier's rank, the harder it was. There was very little intermingling among enlisted men's wives and officers' wives— practically none at all at the purely social level.

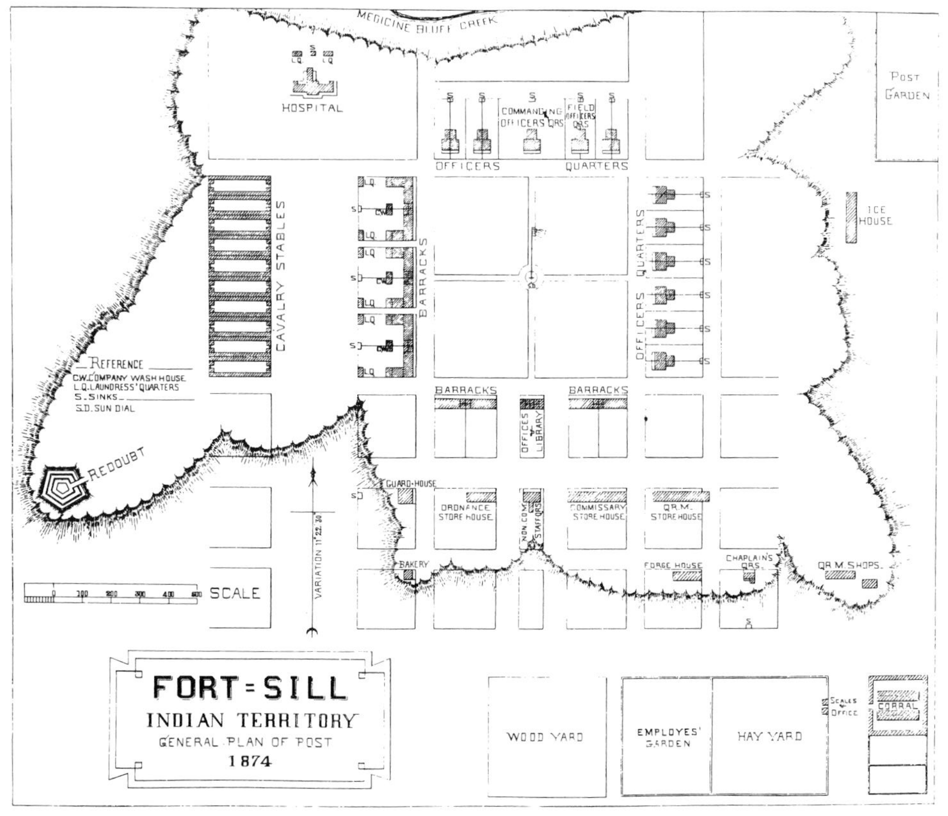

From Circular No. 8, *Report on Hygiene of the U.S. Army.*

Figure 2.1. Quarters of Trumpeter Joseph Nevins, Fort Union, New Mexico, 1886

As noted in the text, there was great variance between the style of living quarters provided for officers in the nineteenth-century army and that for the men under their command. Efforts were made, however, to provide houses or apartments for non-commissioned officers (sergeants and corporals), especially for the men whose wives and families wanted to be with them on post duty, and for married senior enlisted men as well.

In Figure 2.1, Sergeant Joseph Nevins and Mrs. Nevins pose for the camera in their parlor at Fort Union, New Mexico. Mrs. Nevins is busy with needlework. We can guess she has used her sewing skills earlier, to beautify the hassock and the parlor tablecloth; her ornamental touch is obvious in this room, and the desire to decorate according to the prevailing fashion is evident. A lively, figured fabric is used as background trimming for the mantel; neat dust curtains hang at the homemade bookcase on the folding military table at far right. The viewer notices the careful symmetry of the paired pictures hung on the wall, and wonders about the odd "wrap-around" effect of the wall hanging that stretches from a side wall (at left) past the corner onto the rear wall. The decorator's unusual use of fabric in this way may have been simply to add color or interest to the room's drab adobe walls; if so, that purpose was fully accomplished, at least for present-day viewers. Area rugs are carefully placed over the carpet. Clearly, the sergeant's wife has done her best, in a rough environment, to create a proper Victorian parlor and home.

[National Park Service, Fort Union, New Mexico]

Figure 2.2. Sergeant William Minser, 3rd U.S. Infantry, and Friend in the Sergeant's Quarters, Fort Stanton, New Mexico, 1886-1890

Sergeant William Minser, 3rd U.S. Infantry, and a friend enjoy a book in the sergeant's tastefully decorated quarters (Figure 2.2). The quarters provided for non-commissioned officers in the 1800s were usually not so spacious as those for commissioned officers, but they were private quarters. The imaginative noncom could decorate them as he liked, and could entertain guests there, if he wished.

Here, Sergeant Minser entertains his guest by reading aloud, a quite popular pastime of that era, long before radio, movies or television. The sergeant is comfortably seated on what appears to be a bear rug spread over the ingrain carpet on his parlor floor. What looks like bear claws show faintly, just behind the sergeant's right heel. Two handsome platform rockers and a daybed lounge with a second robe or rug spread over the back complete the parlor furniture. Attractive lace curtains appear at the window, and a tasseled fabric is knotted on the back of the rocker at far right. The table cover is a bit of patchwork stitchery that resembles a small "crazy quilt": random pieces of cloth sewn together with bright embroidery thread in bold, jagged ornamental bond, sometimes called "briar-stitching."

[Negative Number 11672. Photograph by Sergeant Charles Harvey, Museum of New Mexico, Santa Fe, New Mexico]

Figure 2.1

Figure 2.2

Figure 2.3

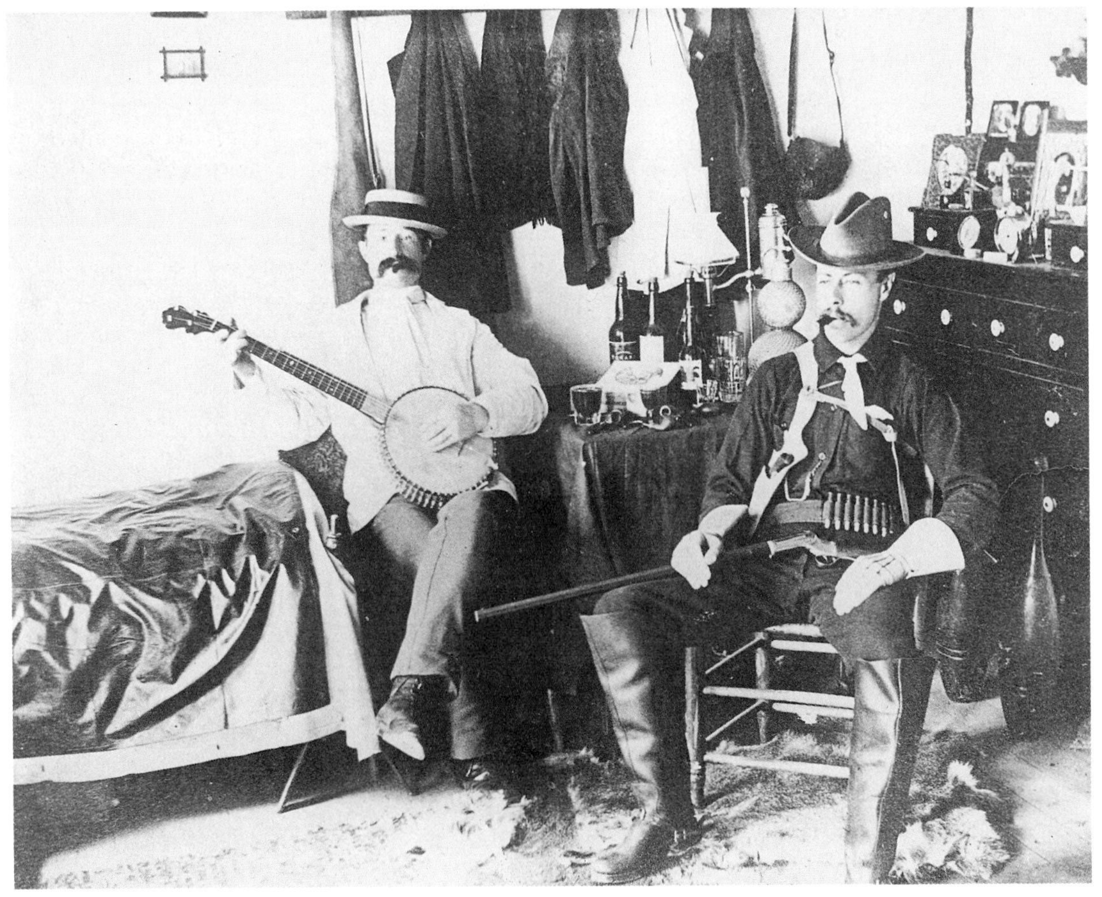

Figure 2.4

Figure 2.3. Sergeant William Minser's Quarters, Fort Stanton, New Mexico, 1885-1890

Sergeant William Minser, seated right, and Sergeant Charles Harvey (at left) concentrate on their chess game in Sergeant Minser's quarters at Fort Stanton, New Mexico. The student lamp at the far side of the game table is typical of the times, as is the center table in the foreground, with books piled high and a reliable kerosene lamp ready for lighting as needed.

The chest of drawers behind Sergeant Harvey was twenty years old or more at the time the picture was made. The large alarm clock seen among the sergeant's memorabilia on the chest was an important working tool for a U.S. Army sergeant, who was responsible for either performing or seeing that others performed scheduled activities at specified times throughout the day. Rustic picture frames like the one on the picture above the chest were popular at this time.

Minser's spare uniforms hang on a peg rack on the wall at right. Next to the rack, his non-commissioned officer's sword, field cap, and a satchel hang from a separate peg. The chair in which Sergeant Minser is sitting is probably of Mexican origin, with a rawhide seat.

[Negative Number 11655, Museum of New Mexico, Santa Fe, New Mexico]

Figure 2.4. Two Noncommissioned Officers, Fort Stanton, New Mexico, 1885

Identifying records for the photograph in Figure 2.4 indicates that the two noncommissioned officers pictured are members of the 3rd U.S. Infantry. I doubt that that identification is correct, however, since the banjo player is wearing cavalry trousers, and the sergeant seated at right has on cavalry boots and gauntlets.

Neither the single-shot, breech-loading rifle laid casually across the lap of the booted man nor the shoulder holster and pistol he wears are military issue. Many of the troops purchased their own weapons for sport, hunting and sometimes even for field use.

The chest of drawers directly behind the booted sergeant is similar to that seen in Figure 2.3. At lower right, directly in front of the chest, are two Indian clubs that the men used for exercise. The table between the two men has a dark cloth cover and supports a student lamp, box of cigars, various whiskey bottles, tumblers, and what appears to be a seltzer bottle of the same kind seen on the chest of drawers in Sergeant Minser's quarters (Figure 2.3). Animal skins spread on the floor serve as a carpet.

The bed at the right of the banjo player is a type of iron bedstead commonly used in army barracks from about 1880. It was called a "Composite Bunk" and was one of two models made by the Composite Iron Works Company of New York and sold to the army from around 1870 until the turn of the century (see Figure 3.20).

A shelter-half (a half-section of canvas from a pup tent) and a rubber blanket have been spread on this bunk as bed covers. Although the army did not provide soldiers with raincoats, it did issue rubber blankets and/or ponchos, both of which were simply pieces of canvas with one side rubberized. Metal grommets, or eyelets, were inserted along the edges of the canvas tent fabric, so that two pieces of canvas could be tied together to form a tent or some sort of lean-to shelter when the men were on field duty. The poncho had an opening in the center, so that it could be pulled over one's head and worn as a raincoat.

[Negative Number 11654. Photograph by Sergeant Charles Harvey, Museum of New Mexico, Santa Fe, New Mexico]

Figure 2.5. Noncommissioned Officers' Quarters, Post unknown, Artillery, 1907

The first object of note is the model 1905 Quartermaster bunk on the left and another on the right. The denim barracks bag is hung from the bunk, exactly the way I did it in basic training at Fort Jackson, South Carolina, in 1958.

The whisk brush hanging near the window is for cleaning his uniform. The object of greatest interest is his dish and the personal effects upon it. The ceramic mug, probably for sewing, marks him as a member of the "Odd Fellows Fraternity." It is a shame that we cannot read the titles on the books in his library. Above the sergeant's head is what appears to be the base of a Model 1881 barracks lamp.

[Jerome Green Collection]

Figure 2.5

ENLISTED MEN'S BARRACKS

"The barracks and quarters of the enlisted men have, as a whole, improved within the last five years," reported the Surgeon General of the U.S. Army on May 1, 1875.[12] He wrote with a well-earned sense of achievement. The office of the Surgeon General of the Army had been campaigning for improvement in the barracks and quarters of the enlisted men at least since 1870, and the great improvement that he refers to in that area was the army's introduction—at long last—of single iron cots or bunks for each enlisted man to sleep on. Prior to the 1870s, enlisted men slept in a "military crib," a stacked up arrangement of, usually, one upper and one lower bunk, with two men sharing each bunk, or four men to a "crib." These "cribs" were usually only two levels high, but sometimes there were three bunks in a stacked unit, and six men slept together.

As early as the 1840s, an Englishman who joined the U.S. Army during the Mexican War (1846-1848) described these sleeping arrangements as "the exceedingly disagreeable custom, still universal in the United States service, of sleeping two men in a bed...a custom which has been abolished in every barracks in Great Britain and the colonies, to the infinite comfort of the soldier, for the last twenty years."[13] It was to be another thirty years, however, and only after tireless campaigning for it by the Surgeon General's Office, before this "disagreeable custom" would cease.

In 1870, the Surgeon General's report spoke forthrightly about this situation.

> There are...other points, in connection with barracks, to which attention should be drawn. Prominent among these, as being a point in which our service is behind the age, and an evil which should be put to an end with the least possible delay, is the use of the double bunk, usually aggravated by placing it in two tiers, and even, as at Fort Buford, in three. These bunks are used in ninety-three, or over one-half our posts. It is certainly time that the use of such bunks should be absolutely and imperatively forbidden, and so long as they are allowed to exist in dormitories, so long is it useless to hope that those rooms can be made what they should be. No one acquainted with the first principles of sanitary science will approve of their use.[14]

Up to the 1870s, the army's view had been that the common soldier was generally on duty out in the field, where he was supposed to live in a tent and out of his knapsack. Army policy-makers felt that the service was being more than generous in providing the enlisted men sleeping quarters that had a roof and the "military crib." Use of single cots per man was reserved for the post hospital and for cadets at West Point. The Surgeon General continued to lead the fight for improved lining conditions for the enlisted men, however, and his success may be judged by the changes brought about in 1875. His report in May of that year continues:

> I am glad to say that the double and two-story wooden bunks are now nearly abolished, and that the iron bunks now furnished by the Quartermaster's Department are very satisfactory, with the exception of a few, which are two-story in pattern—that is, an iron frame containing two beds, one four or five feet above

> the other. Under no circumstances, except for the most temporary emergency, should beds be arranged in this manner. It is connected with the deficient air-space, and gives an appearance of room where there is not. Every man should have his sixty square feet of floor space, as much as his ration.
>
> But even with single bunks the supply of bedding is unsatisfactory. No sheets or pillows are furnished, and the men come into direct contact with the blankets, and use their greatcoats for pillows. The blankets are seldom washed, although they are aired and beaten occasionally. The bed-sacks are usually too short, and, as Colonel C.H. Smith, nineteenth United States Infantry remarks, "No amount of too short bed can make a man comfortable."
>
> The recommendation of Dr. Pitzki, that wire mattresses, hair pillows, and sheets be furnished for the troops, is believed to be a good one, the results which, in promoting comfort and content among the men, would be a full equivalent for the money it would cost.[15]

Since pictures of barracks in which the "military cribs" were used are quite rare, a number of photographs of the few that survived the 1875 change-over are included in this book: see photographs and legends for Figures 3.1 through 3.7.

And, throughout this section of the book especially, the reader's attention is directed to specific aspects of each photograph that show the effort made by the Office of the Quartermaster General to provide barracks that met the standards in all army barracks, whether they were at posts near large cities or in stockades surrounded by sand and sagebrush and—most particularly—whether they housed men who were black, white, or red, and the nineteenth-century army had soldiers in each category; see photographs and legends for Figures 3.25 through 3.28. The racial segregation that existed in the U.S. armed forces from very early times until after World War II (see legend, Figure 3.25) made no noticeable difference to the army's Quartermaster Corps. A soldier was a soldier, whatever his color, and barracks furnishings for all enlisted men were as nearly the same as could be achieved, throughout the system.

Most of the photographs in this section are from the National Archives, records of the Office of the Quartermaster General. These photographs were taken between 1891 and 1895, as part of a major survey of existing army posts of the times, and most of them show exterior views only of the army buildings in the survey. Frequently, however, the photographers also shot interior views of the post barracks. Few other pictures of the inside of army post buildings of that era were made.

Readers who served in the U.S. Army before the 1960s will recognize details in some photographs here that are similar to the squad rooms they themselves knew, in service. Since the 1960s, however, the army has been moving away from the squad bay or squad room, and most enlisted personnel are now billeted in rooms housing two to four soldiers.

This major change has moved us a giant step away from the army of the nineteenth century. Gone are the endless personal adjustments each soldier once needed to make, in the old days, so that forty to sixty men could live in one room without annihilating each other. Gone is a total lack of privacy, the inevitable two or three men who persistently snored; also gone and unlamented is the oppressively foggy atmosphere of too many warm bodies housed in a space too small and poorly ventilated. Before the daily bath became routine for most Americans, the close and heavily laden air in a nineteenth-century army barracks must have been overwhelming, especially in July and August.

Figures 3.1: A, B, C. Half of a military bunk (crib), at Fort Mifflin, Pa., ca. 1800-1830

This is the earliest surviving military bunk that is known. It is the top half of a double bunk. The holes that appear along the side boards and end boards were done in the 1970s by some misguided fellows who tried to make this a rope bed.

In photo A, the rail used to support boards running across the bed can be seen. It should be noted that the bunks ca. 1840 also have this arrangement for supporting the mattress.

This bunk does not "knock apart" as does the later bunks. This feature dates the bunk prior to 1830. The wood appears to be cedar and has many coats of white-wash on it. The bunk measures about 42 wide and 70 long.

[The personal collection of the author]

Figure 3.1 A

Figure 3.1 B

Figure 3.1 C

Figures 3.2: A, B, C, D. U.S. Army Military Crib, Fort Scott, Kansas, 1840

The wooden framework and slats of the U.S. Army's ''military crib''—the nineteenth-century enlisted man's bed—are shown in Figures A through D. The crib photographed here is used at Fort Scott, Kansas. Its form, construction, and the walnut wood it was made of date this bunk at about 1830 to 1840. It measures four feet wide by six feet long, consists of two wide bunks, one upper and one lower, and was built to hold four soldiers, two to a bunk.

Each bunk was equipped with a mattress made of bed ticking filled with straw, and each soldier had one or two woolen blankets. No sheets or pillows were provided. The men often folded their overcoats and used them as pillows.

Mortise-and-tenon construction held together by reinforcing wedges of wood driven into strategic holes in the wooden frame (see Figures B and C) made the bunks easy to take apart—or knock apart, as the troops termed it—for thorough cleaning. And periodic thorough cleaning of the bunks was essential in nineteenth-century military barracks, because the soldiers' beds were chronically infested by that era's common parasitic vermin—bedbugs, body lice, and fleas. Once these pests established themselves in crowded barracks, they were extremely hard to get rid of. Bunks were, therefore, routinely moved outside, spring and fall, and taken apart to be scrubbed, scoured and whitewashed. While the bunks were outside, the barracks floors were sanded, and the room's walls and ceiling were also whitewashed.

Bunks of the type shown in Figures A through D did not have a storage rack or a shelf at the end of the frame, which probably indicates that the rooms in which they were used were provided with wall shelves with pegs.

Originally from the Philadelphia area, the military crib photographed here is now in the collection of Fort Scott National Historic Site, Kansas.

[Fort Scott National Historic Site, Kansas. Photographs by the author]

Figures 3.3: A and B. U.S. Army Military Crib, Fort Mifflin, Pennsylvania, 1861

The bunk framework shown in Figures A and B is one of nine surviving U.S. Army cribs made in 1861, when Fort Mifflin was regarrisoned for service during the Civil War. These four-man bunks were also used for Confederate prisoners.

On earlier army bunks, the wooden slats that supported the mattress ran crosswise of the frame, from side to side; but in this model, they ran lengthwise, from headboard to footboard. These long slats were held in place by the frame's wide end boards, which were nailed to the uprights (see Figure B). As was true to the crib seen in Figures 3.2 (A) through (D), the beds at Fort Mifflin could be taken apart by removing the wedges (Figure 3.3 B) that tightened and held the framework together. The wooden frames of the nine Fort Mifflin cribs all have large surface areas where the whitewash brushed on during seasonal cleanings still clings.

[Fort Mifflin, Pennsylvania. Photographs by the author]

Figure 3.2 A

Figure 3.2 B

Figure 3.2 C

Figure 3.2 D

Figure 3.3 A

Figure 3.3 B

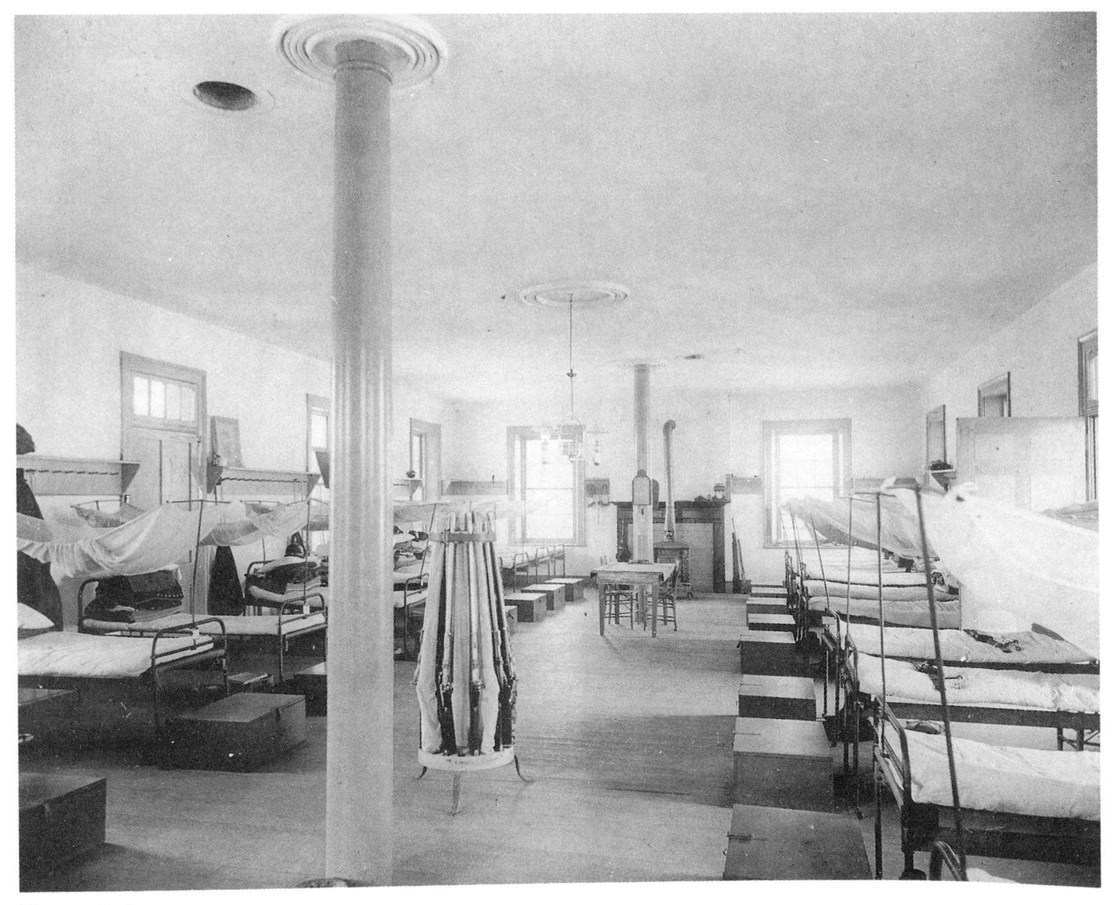

Figure 3.4

Figure 3.5

Figure 3.4. Old Infantry Barracks, Jackson Barracks, Louisiana, 1891-1895

This photograph is included because it shows a barracks room with wall shelving for storage on each of the three walls visible. The fourth wall (behind the photographer when the picture was made) also has these storage shelves. If we could remove from this room the single rack at the center foreground and the two-burner lamp suspended from the ceiling, and in their places put the military cribs shown in Figures 3.2 through 3.3, we should see the sort of U.S. Army barracks in use before the Civil War. The rifle rack seen here in center foreground would be altogether different, as would the lighting devices. For lighting, the pre-Civil War army provided candles—but very few of them.

[Negative Number 92-F-28-5, National Archives, Washington, D.C.]

Figure 3.5. "Stag Dance," *Harper's Weekly,* February 6, 1864

This barracks is identified as temporary winter quarters for the troops enjoying a bit of free recreational activity inside it, out of the weather. Come spring, these soldiers would be back outside in tents. The artist who recorded this scene for *Harper's Weekly* shows us the commonly used two level double bunks, or cribs, which provided sleeping space, of a sort, for four men, two above and two below. In most barracks buildings of this period, the bunks would be free-standing, but here they are build right into the wall. A plan of Fort William Henry, at Lake George, New York from approximately 1755, shows bunks arranged just like this.

[*Harper's Weekly,* February 6, 1864]

Figure 3.6. Temporary Barracks, Washington, D.C., 1861-1863

Figure 3.6 is the only photograph I know of that shows the "military crib" in full use during its heyday. This view is one-half of a stereoscopic pair. The bunks it shows are the three-tiered kind, which were often used in temporary army barracks. The bunks in Figure 3.6 appear to be secured against the walls, as were the bunks in Figure 3.5. The total lack of barracks furniture is immediately noticeable; not a chair, bench, table, or lighting device can be seen anywhere in the room. Its absence was not a matter of its having been pushed out of the way to clear the scene for the photographer either; this barracks, like many other temporary army barracks, simply did not provide such comforts as furniture, since none of the men billeted in a temporary barracks stayed there very long. Generally speaking, regiments assigned temporary quarters were on the move and were quickly transferred from this sort of housing into the field, where they were sheltered under canvas.

Viewers with sharp eyes may notice in the right foreground, just forward of the bunk ladder, the tiny white cards affixed to the side rails of the bunks. These cards carried the name and number of the soldier to whom the bunk was assigned. Each soldier's uniform, equipment, and weapon also carried his name and number, so that, during an inspection, it was a simple matter to check who had what and what condition it was in.

[Mike McAfee Collection, West Point, New York]

Figure 3.7. Infantry Barracks, 1st Regiment, Pennsylvania National Guard, 1895-1898

When this picture was made, at the tag-end of the nineteenth century, the U.S. Army had come to consider the military crib as sleeping quarters completely out of date. This barracks did not house regular army men, however; it housed a National Guard unit, and in the peacetime years of the later nineteenth century, the National Guard was made up entirely of state-governed and state-maintained units throughout the country. Until such time as a national emergency should arise, there was no formal connection between the state-governed National Guard units and the federal government. In that way—and many others—the National Guard was much like the old-time state militia units. The governor of each state was commander-in-chief of the guardsmen in his state and each state government was responsible for training, equipping, and maintaining the National Guard units within their boundaries. The various states more or less followed U.S. Army regulations, in general, often adapting the rules here and there, as economics and official leniency dictated. Peacetime National Guard units all over the country were, therefore, trained, uniformed, drilled, and supported with widely varying degrees of efficiency from state to state. For these reasons, it is likely that much of the equipment issued to any unit of the National Guardsmen in the 1890s would be as much out of date as the military crib was.

One notices, in Figure 3.7, a storage shelf at the end of the near lower bunk and what appears to be small storage cabinets between the upper bunks and the wall. While there were no specific regulations concerning the way a military crib should be built, much can be inferred from army inspection reports of the pre-Civil War era. Usually, the run-of-the-mill four-man bunk was a double bunk that could be taken apart for cleaning. Ordinarily, it had either a rack at one end, for the storage of weapons, equipment, and clothing, or a wall shelf, with separate gun racks nearby. In a way, it seems odd to us that the army should not have had clear drawings and specific directions for the construction of items like this, but such well-organized planning did not come about until after the Civil War. As a matter of fact, as early as 1826, the Inspector General, Colonel George Croghan, wrote in an inspection report: "It is necessary that correct drawings of both bunks and arms racks, exhibiting their forms, position with relation to the chamber, mode of numbering, etc., be furnished to each post.[16]

Unfortunately, Colonel Croghan's suggestion was not followed. As Inspector General of the army from 1826 to 1845, the colonel examined all frontier army posts. His reports provide a wealth of information on army life during this period.

Throughout the extent of its use as sleeping quarters for the soldiers, about all that could be said for the military crib is that these units were either two or three tiers in height, generally of the same style and construction, and normally could be knocked apart for cleaning. [103rd Engineers, Battalion Museum, Philadelphia, Pennsylvania]

Figure 3.6

Figure 3.7

Figure 3.8

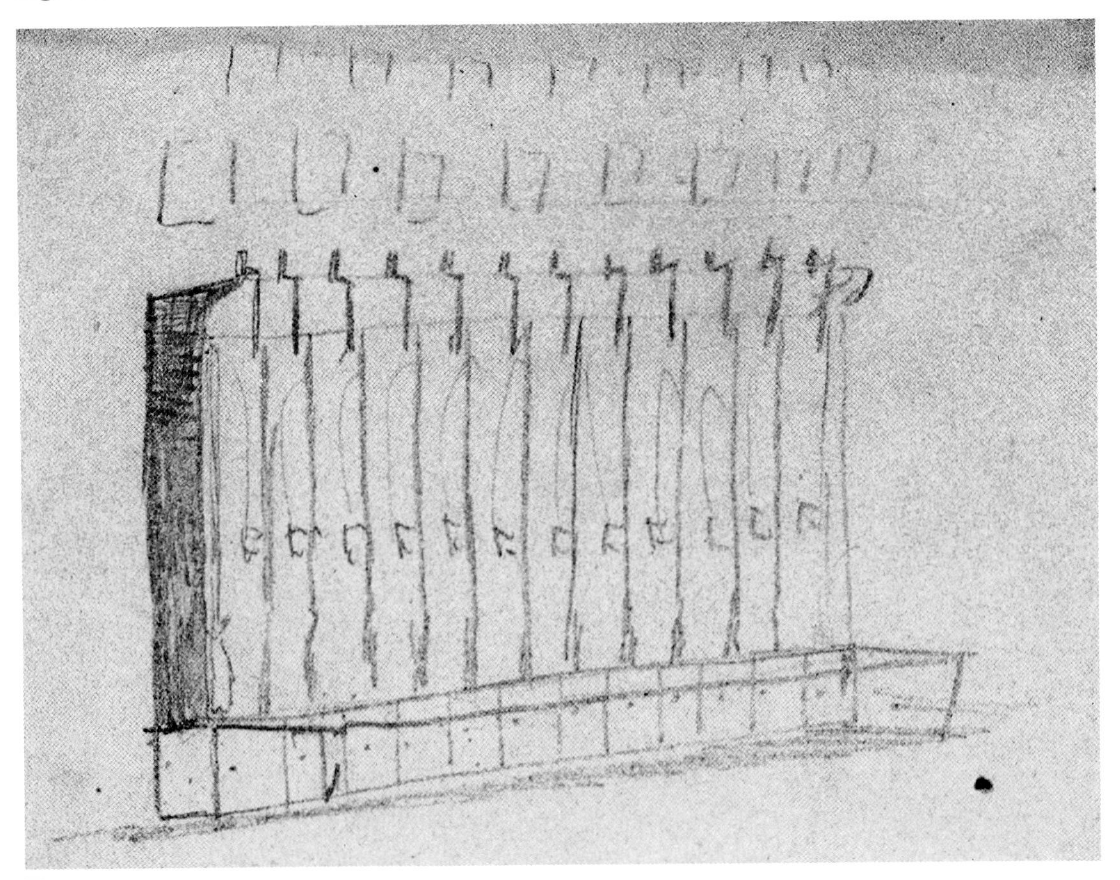

Figure 3.9

Figure 3.8. Temporary Sleeping Quarters in the U.S. Patent Office, Washington, D.C., for the Rhode Island Regiment, 1861

Early in the Civil War, the Rhode Island Regiment of the U.S. Army was housed in temporary barracks in the U.S. Patent Office in Washington, D.C. Lacking space for footlockers or shelves in a room never intended to be slept in, the men adapted to the situation and hung clothing and equipment from the bunkposts. They did have storage space for their rifles; arms racks were conveniently built into the bunk-ends.

In 1864, the U.S. Quartermaster General's Office issued a plan (General Order No. 17, National Archives, Washington, D.C.) calling for three-tiered bunks of the same kind as those seen here, to be used in all temporary barracks.

[*Harper's Weekly,* July 18, 1861]

Figure 3.9. Arms Rack (Sketched, ca. 1846)

The roughly sketched arms and equipment rack in Figure 3.9 was found drawn on the inside cover of an infantry tactics manual printed in Philadelphia in 1846. The sketch shows what appear to be lockers built in at the rack's base; the musket rack is just above the lockers. Cartridge boxes on slings stand behind each weapon, and the bayonets are secured over a bar across the unit's top. The men's dress helmets may have been stored in the spaces sketched in above the bayonets. It should be noted here that, while we do not have an actual, accurate sketch of a pre-1870s army weapons rack, we do know from information in many inspection reports, that most such racks were free-standing and circular in form.

[J. Craig Nanos Collection, Upper Darby Pennsylvania]

WAR DEPARTMENT,

QUARTERMASTER GENERAL'S OFFICE.

Specifications for Mattress-covers.

Duck.—To be made of cotton duck, forty (40) inches wide, weighing from eight and three-quarters (8¾) to nine (9) ounces to the linear yard.

To have fifty-three (53) threads of two-ply yarn to the inch of warp and thirty-four (34) threads of single yarn to the inch of filling.

To stand a breaking strain of not less than seventy (70) pounds to the half-inch of warp and not less than twenty-one (21) pounds to the half-inch of filling.

To be entirely free from sizing.

Dimensions.—Length when finished seventy-eight (78) inches. Width when finished thirty-nine (39) inches. The covers to have box corners four (4) inches deep.

Making.—The side seams to be felled down. The open end to have a two (2) inch hem, when finished. The top and bottom to overlap each other about one and a half (1½) inches at open end and be fastened by six (6) tapes, three-quarters (¾) of an inch wide and twelve (12) inches long, placed opposite each other on top and bottom sides, about nine (9) inches apart.

The covers, after being made up, to be smoothly pressed.

Materials, workmanship, and finish.—To be like and equal in all respects to the sealed standard sample.

Adopted January 20, 1886.

S. B. HOLABIRD,

Quartermaster General, U. S. A.

128—F., 1886.

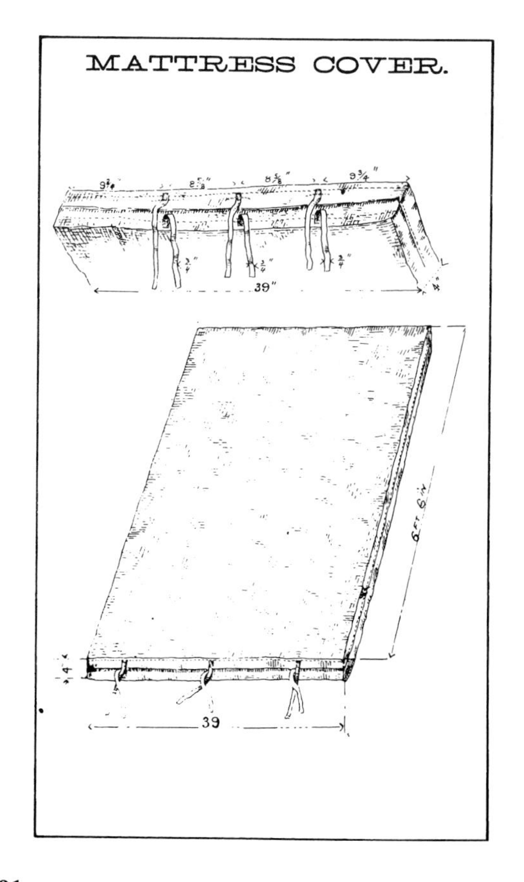

Figure 3.10. Infantry Barracks, Company A, Fort Leavenworth, Kansas, 1874

Figure 3.10 is one of the earliest known photographic images of a U.S. Army barracks interior. This photographic "first" is additionally important because it shows the beginning of transition in furnishing army barracks interiors during the latter part of the 1800s.

When this photograph was made, the army had recently purchased more than four thousand bunks of the type shown here, built by Snead and Company of Louisville, Kentucky. Sharp-eyed viewers will notice on the first bunk (right foreground) a distinctive point about Snead bunks: the wooden bed slats used to support the mattresses did not run from side to side of the bunk frame, as do most bed slats we are accustomed to; they ran lengthwise, from headboard to footboard of the bunk. Close up views of the Snead bunk appear in Figures 3.11 (A) through (C).

The footlockers at the foot of several bunks were not regulation army equipment for that time and would not be for some years to come. The small wooden box on the floor at center foreground is the troopers' "spit box"—a homemade spittoon that would have been kept half-filled with sand. What appears to be a tin candle sconce dimly seen on the back wall behind the stove, at left, was a non-regulation lighting device and would have been purchased with company funds.

Dominating the picture (near right foreground and center) are the two round arms racks fixed to the room's supporting columns and, between them, the early fire-fighting equipment on a heavy free-standing round stand. The cylindrical object fixed at the top of the stand holding the leather fire buckets is a Babcock Patent fire extinguisher (Records of the Patent Office, RG-241, National Archives Cartographic Center, Washington, D.C.). Fire axes and two speaking trumpets are secured on the stand below the extinguisher, with buckets hanging from the lowest—and largest—rack. The wooden barrel with the letters "A," to the right of the fire-fighting equipment, is a water cooler. The "A" is the company's letter-designation.

[Fort Leavenworth National Historic Site, 6003.46-C]

Figures 3.11: A, B, C. Barracks Bunks, Fort Larned National Historic Site, Kansas, 1870

Between November 1871 and June 1872, the U.S. Army bought from Snead and Company, of Louisville, Kentucky, more than four thousand wrought-iron barracks bunks like the one in Figure B. The Snead bunk (also seen in Figure 3.10) was early standard army barracks equipment. It was plain, sturdy, and reliable, and—as noted earlier—its wooden slats ran lengthwise of the frame, from headboard to footboard. The boards shown in use as bed slats on the model photographed here may be the original slats issued, although they look a bit thick when compared to the Quartermaster's drawings and the bunk in Figure 3.10. The bunk that served as the photographer's model for the three shots in this series survives today at Fort Larned National Historic Site.

[Fort Larned National Historic Site, Kansas. Photographs by the author]

Figure 3.10

Figure 3.11 A

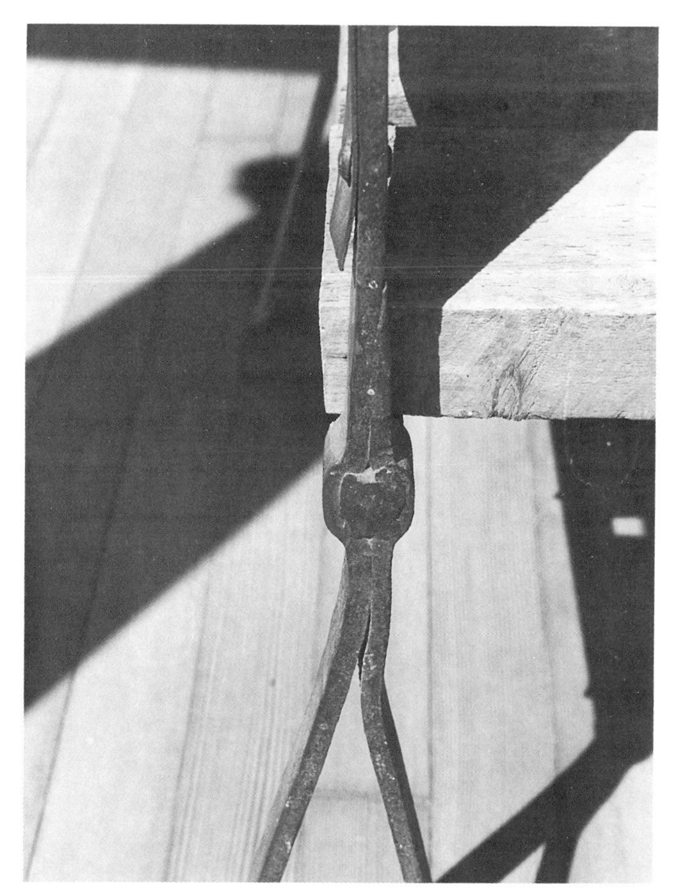

Figure 3.11 C

Figure 3.11 B

Figure 3.12

WAR DEPARTMENT,

QUARTERMASTER GENERAL'S OFFICE.

Specifications for Mosquito Bars.

Material.—To be made of the best quality barred mosquito netting and white cotton tape, equal in quality to the same materials in the standard sample.

Dimensions.—Seven (7) feet long, two (2) feet eight (8) inches wide, and five (5) feet eight (8) inches high.

To be bound around top and down the four corners with white tape, and to have two (2) strings of white tape nine (9) inches long, strongly sewed on each of the four upper corners, and to conform in all respects to the standard sample adopted May 23, 1884.

Adopted June 7, 1884.

S. B. HOLABIRD.

Quartermaster General, U. S. A.

1250—F., 1884.

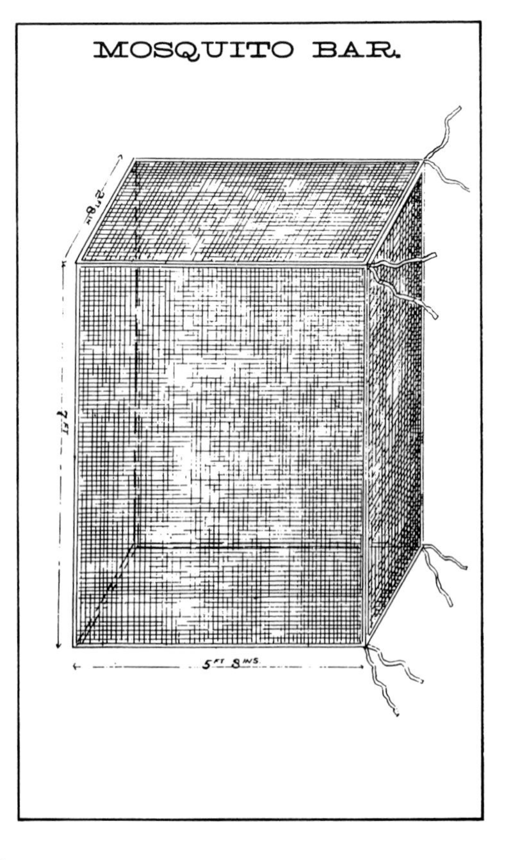

Figure 3.12. Infantry Barracks, Fort Leavenworth, Kansas, 1888

Progress is seen in the barracks bunks of the late 1880s; in Figure 3.12, the old barracks bunks by Snead and Company of Louisville, Kentucky offer something new for the soldiers' comfort. Instead of the wooden boards used in earlier days as bed slats to support bedding and sleeper, here is an early type of metal bedspring in the form of a woven-wire pad fixed to each end of the bunk frame. The woven-wire supports are first mentioned in the annual Quartermaster's report for 1884.

Soldiers' spare clothing and equipment hang from the pegs of the storage rack on the wall behind the bunks. One notes that the soldiers turned their overcoats inside-out before hanging them up, and, of course, every individual who has ever worn a U.S. Army uniform will recognize the regulation denim barracks bags issued to every soldier.

[Negative Number RG-92, CCFGM, National Archives, Washington, D.C.]

WAR DEPARTMENT,

QUARTERMASTER GENERAL'S OFFICE.

Specifications for Pillows.

To conform in all respects to the sealed standard sample.

Materials.—To be made of narrow striped blue and white ticking, "herring-bone" or "twill" weave: the filling to be of good cotton linters.

Dimensions.—To be thirty (30) inches long and seventeen (17) inches wide, and to weigh not less than three (3) pounds.

Workmanship.—To be made in a neat and substantial manner.

Adopted March 28, 1885.

S. B. HOLABIRD,
Quartermaster General, U. S. A.

1075—F., 1885.

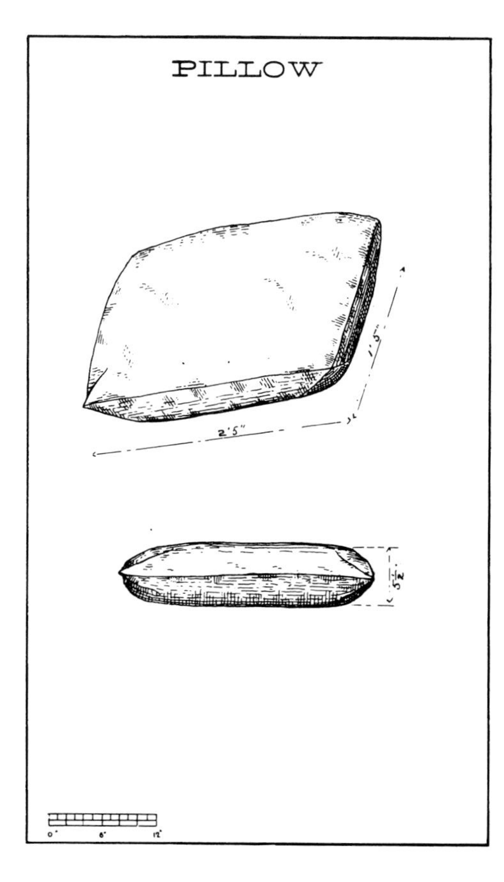

Figures 3.13: A, B. Floor Plan for Two-Story Barracks Recommended to the Secretary of War, September, 1872

The floor plan in Figures 3.13 A and B for a two-story army barracks was recommended to the Secretary of War by Quartermaster General Montgomery Meigs in his report of September 14, 1872. An excellent example of an army barracks built following this plan survives at Fort Laramie National Historic Site, Wyoming. This carefully drawn plan should be a helpful reference for viewers looking at various photographs that follow in this section as well as the later photographs of army mess halls and kitchens.

As the plan indicates, the barracks sleeping area was to be on the building's second floor, and some sort of "window treatment"—shades, shutters, curtains, draperies, or other movable means of covering the windows—would be required to provide a measure of privacy for the men inside. It should be kept in mind that window treatments in nineteenth-century army barracks were not universally provided, by any means. In the very close social environment in which army posts existed, if the men's sleeping area was on the first floor of a barracks building, window curtains or something similar were required. When the sleeping area was on the second floor of the barracks, however, window coverings had not, until then, been considered essential.

In the interest of historical integrity, I should state here that in some of the photographs that follow, a good many of the coverings on barracks windows have been placed there by the photographer to control the light inside the room while the interior was being photographed.

[Number 33462-RN, U.S. 273.A, 1882, from the Collections of the Library of Congress, Washington, D.C.]

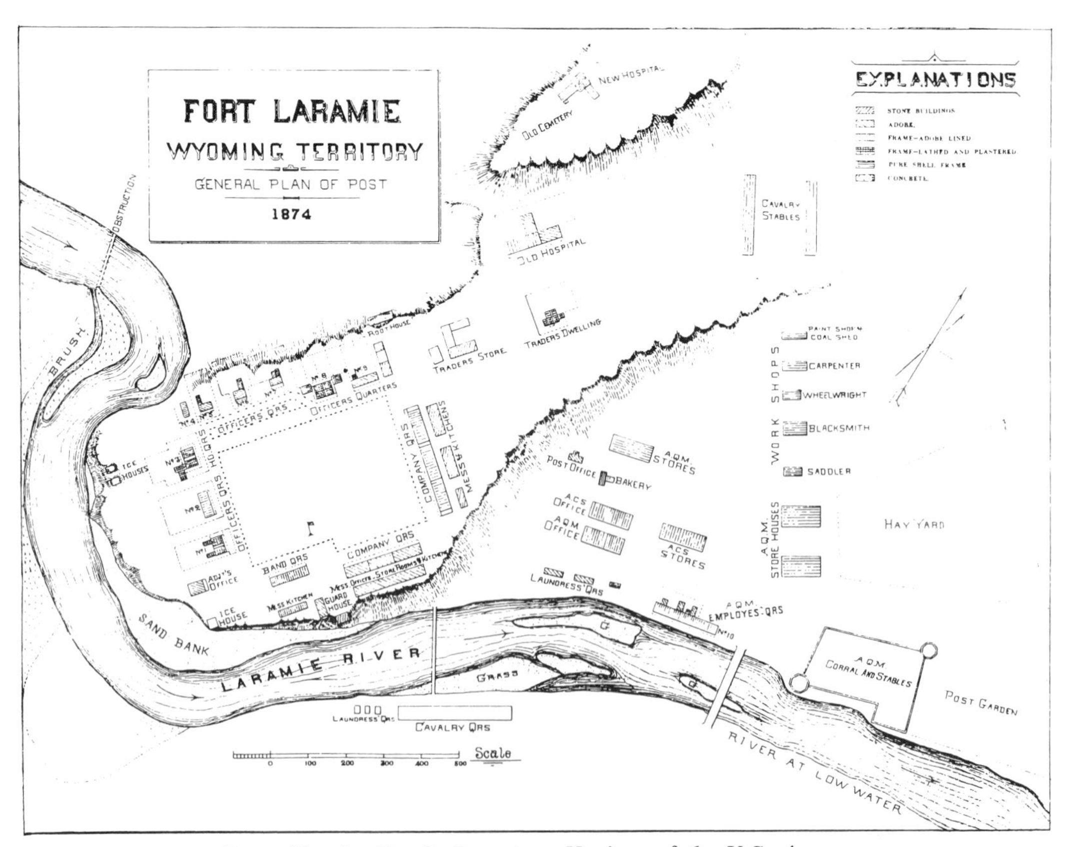

From Circular No. 8, *Report on Hygiene of the U.S. Army.*

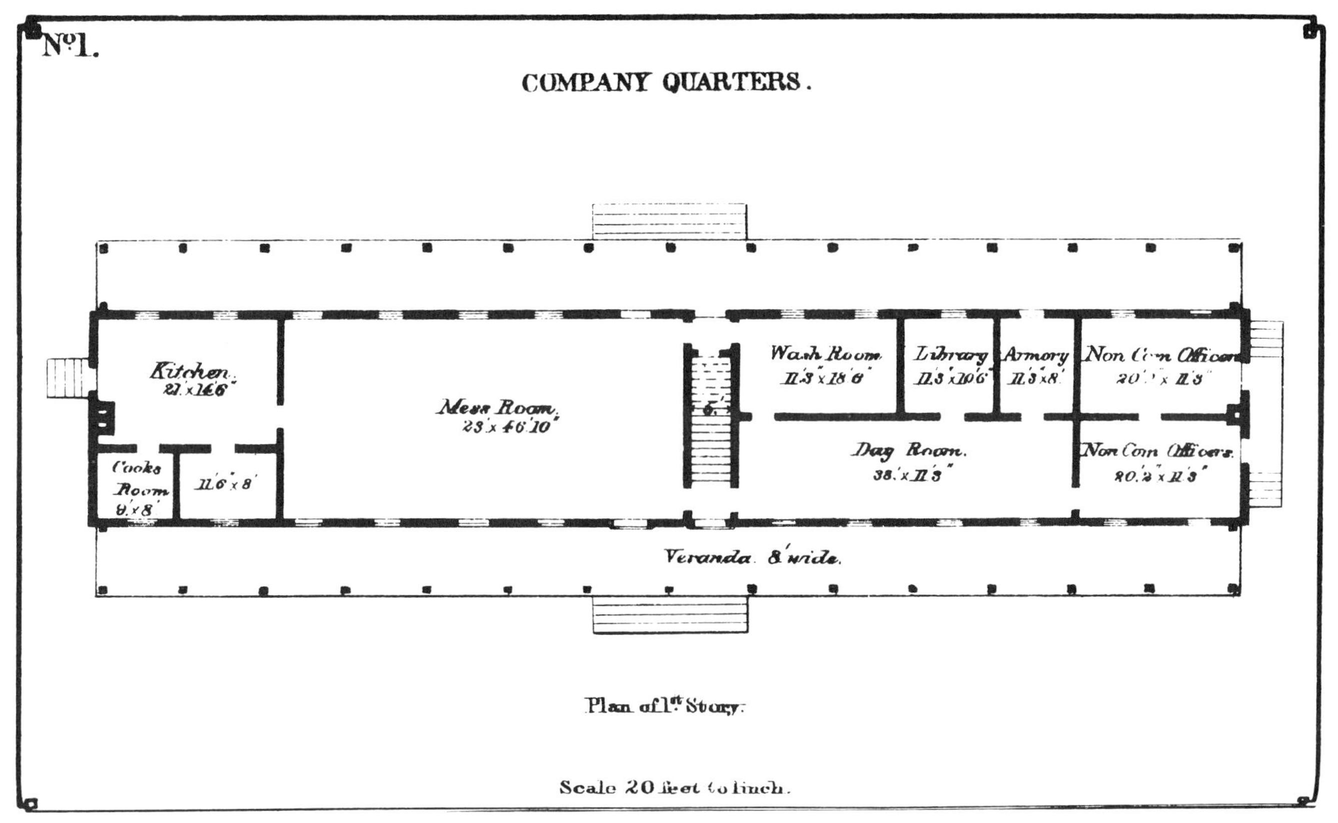

Figure 3.13 A

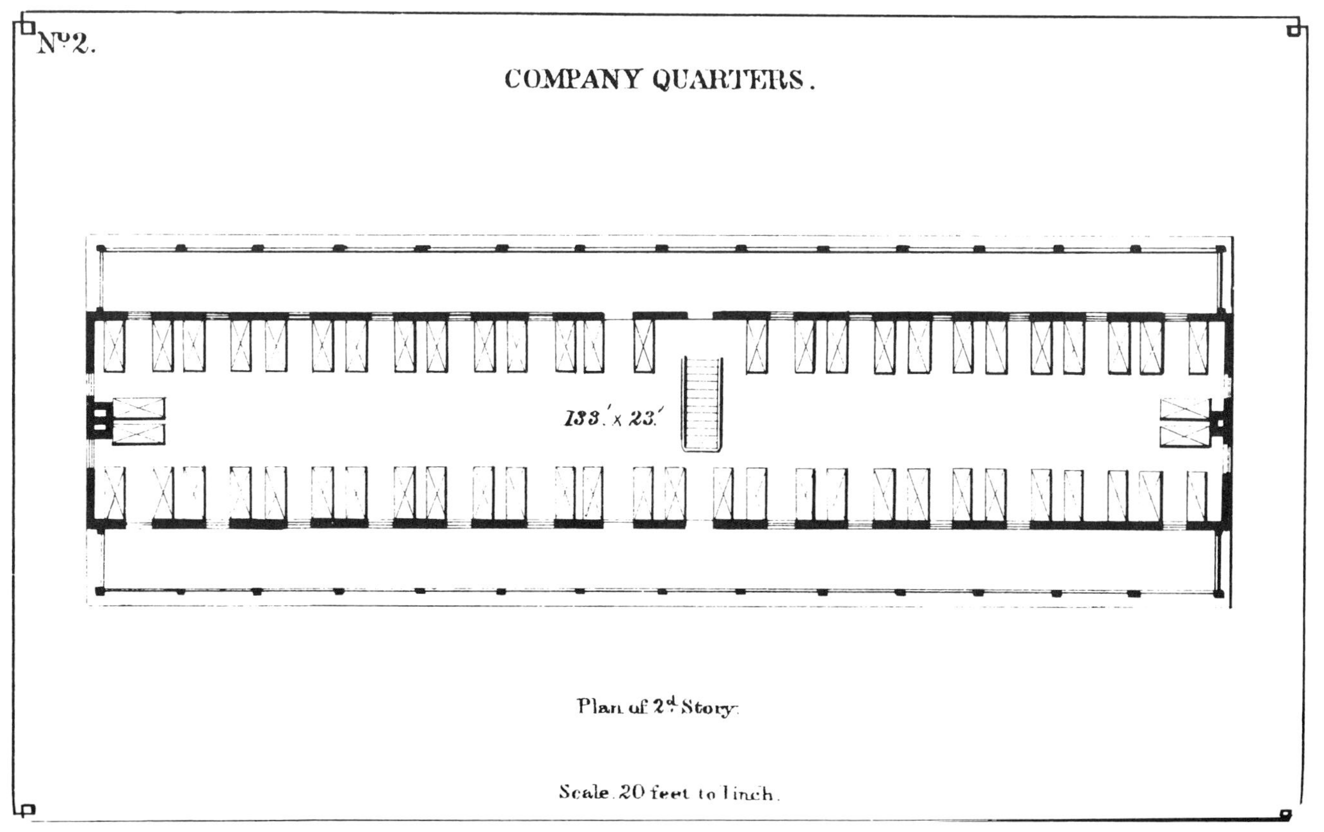

Figure 3.13 B

Figure 3.14. Infantry Barracks, Angel Island, California, 1891-1895

The typical 1890s army barracks sleeping area shown in Figure 3.14 was at the Angel Island, California, infantry barracks. The Angel Island barracks building was constructed according to the Quartermaster General's 1872 plan for a two-story building (see Figures 3.13 A and B), with the sleeping area on the second floor.

The balustrade in the middle of the room surrounds the staircase linking the first and second floors. On the left-hand wall, near the ceiling, at about the spot where the stairway ends, is a large wall clock, its face and case barely discernible because of the acute angle at which it was photographed.

The wall shelving that runs completely around the room, between door and window frames, is almost an anachronism this near the turn of the century; wall lockers were already being used extensively by the 1890s.

At upper right is the army's "Model 1881 Army Lamp, Two-Burner, Pendant." Attached to its top—and to that of its twin, just visible at the far end of this long room—are rounded smoke shields made of what appears to be thin sheet metal. In noting the meager number of lamps in this room, one also notes the logical placement of the barracks tables and chairs used for socializing, reading writing, and checker-playing. They are centered directly beneath the room's only two sources of artificial light.

The footlockers, like those seen earlier in Figure 3.10 are not regulation style. The bunks have been fitted with the woven-wire mattress supports seen earlier in Figure 3.11 (C); in the immediate left foreground, at the head of the second bunk near the wall, we can see the mattress of striped bed ticking, with a soldier's blankets and bedroll stacked on top of it. The tin washbasin under the bunk of that same soldier indicates that the average enlisted man's personal property included at least basic equipment for personal cleanliness.

While this photograph was being made, the rear windows of the room were covered with blankets to control the light. In the immediate right foreground, resting atop the first footlocker in the row, we see the photographer's extra lens and shutter bag.

[Negative Number 92-F-3-4, National Archives, Washington, D.C.]

Figure 3.15. Infantry Barracks, Bencia, California, 1891-1895

The two different models of the same kind of bunk seen at left and right along the walls of the barracks room in Figure 3.15 were both made by the Composite Iron Works Company of New York. This type of bunk, with its simple wrought-iron frame—surprisingly decorative—was by far the most common type of army bunk in use from the late 1870s until 1890. Between November 1871 and June 1876, the army purchased a number of this company's earlier models, the Composite Bunk No. 9. Bunk No. 9 is the one whose tall headboard and footboard extend upward a foot or so above the mattress and bedding (see bunks 2,3,4, and 5 in the row at left in Figure 3.15).

In 1880, the army began buying the company's new model, the Composite Bunk No. 10, and continued buying the No. 10 through the early 1890s. This bunk has much shorter, less cumbersome foot and headboards than the earlier No. 9. The first bunk in the row at left in Figure 3.21 and its opposite number in the row on the right side of the photograph are Composite's Bunk No. 10.

The footlockers in this dormitory conform to the 1875 army regulations as according to General Orders and Circulars, No. 56, Adjutant General's Office, April 30, 1875 (Washington, D.C., Government Printing Office, 1876).

The three circular arms racks seen here are a type mentioned in Ordnance Note No. 125 in the 1880 Report of the Chief of Ordnance. These racks held twenty rifles. Army lamp Model 1881 is clearly visible at upper center, and centered at the far end of the room is the stove, the army's "Cast-Iron Coal Heater VI," first mentioned in specifications dated 1875.

Grace notes in this military room include the dust covers placed over the wall shelves and numerous framed photographs visible on the walls.

[Negative Number RC-92-F-8-4, National Archives, Washington, D.C.]

Figure 3.14

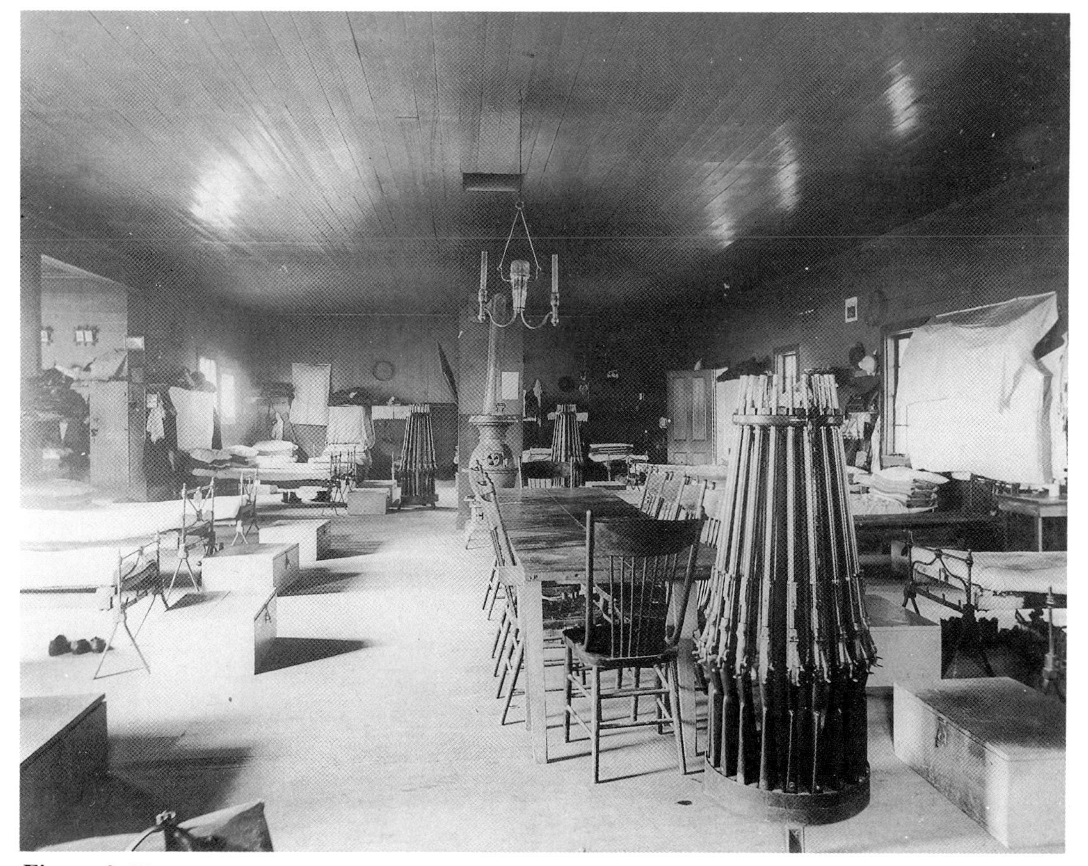

Figure 3.15

Figure 3.16

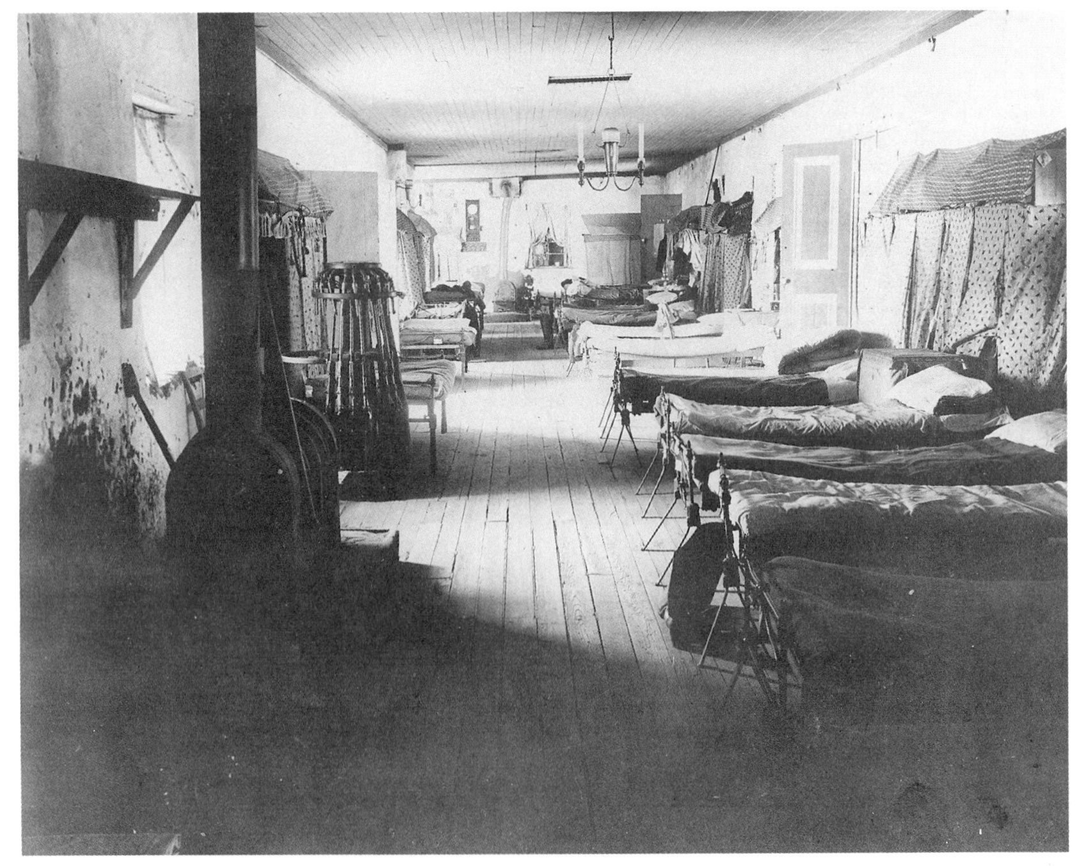
Figure 3.17

Figures 3.16 and 3.17. Cavalry Barracks, Exterior and Interior, Fort Grant, Arizona, 1891-1895

Often, the army did not proceed strictly according to regulations in constructing the buildings an army post needed. Native materials were often used, and what an army building might be made of frequently depended on what section of the country it was to be built in and which materials were readily available there. In Figure 3.16, we see the outside of a one-story adobe barracks building. The fort that needed this building was in Arizona where lumber for building was both costly and less easily obtained than materials closer at hand. Adobe bricks, however, could be had in Arizona with no trouble at all, and they could be made into quite durable buildings. Fort Grant's cavalry barracks was such a building. It was crowded, however, and very dusty; the situation was much discussed by the Surgeon General, who did not approve of any part of it, particularly the adobe bricks that it was made of—an interesting sidelight, since the word "barracks," comes from the Spanish work *barraca* (cabin, mud hut), which in turn comes from the Latin word *barrum* (clay).

Inside the Fort Grant Cavalry Barracks (Figure 3.17), we see two army lamps, one a Model 1881 two-burner pendant lamp (upper center) and the other, only faintly visible at left on the rear wall, a single-burner Model 1881 bracket lamp.

At left foreground, the arms rack, with an upper shelf for pistols and a lower one for carbines, identifies this room as a cavalry barracks.

The variety of cheap, printed-cotton fabrics used for dust covers and as a scarf on the clock shelf at the far end of the room is an interesting facet of the decor in this military building. At the far side of the fourth bunk (right foreground), a civilian trunk is used as a bedside stand. Continuing down the right side of the room toward the back, we see the tall glass chimney of a kerosene lamp, at about the eighth bunk. Such a lamp would have been privately owned, and kerosene to fill it could have been purchased at the post's quartermaster stores or from the post sutler.

[Negative Numbers RG-92-F-24-1 and RG-92-F-24-2, National Archives, Washington, D.C.]

Figure 3.18. Cavalry Barracks, Fort Leavenworth, Kansas, 1891-1895

Three types of army bunks used during the last quarter of the 1800s are seen in Figure 3.18. The first three bunks in the center foreground are the Composite Iron Works Company's Bunk No. 9—the one with the tall footboards. Across the aisle, on the right side of the room, is the newer Bunk No. 10, with the shorter footboard. And, on the extreme right, is a Snead Barracks Bunk, identifiable by its legs and its plain metal footboard.

The viewer will have noticed that, in some army barracks, the mattresses and blankets are folded or rolled up and stacked on top of the bed's bare slats or woven-sire springs during the day, and in other barracks, the bedclothes are simply spread up neatly and left in place on the bunk, ready to be slept on the following night. The procedure followed would vary with each different army post. Fort Leavenworth, at this particular period, followed the practice of simply "making up" the beds with the bedclothes left in place. The bunk in the foreground here is the one with mattress and blankets folded and stacked and the springs left bare. That treatment for a single bunk or a whole room full would indicate that the trooper who usually slept there was either out on guard duty or in the hospital.

The two barracks lamps, at the upper right and left center at the far end of the room, are non-regulation, probably purchased out of company funds. Although it would be hard to discover them without a cure from the photo file, this room was heated by radiators. One of them can be seen, just barely, at the center of the photo, where it is currently in use as a prop for the second man from the left. A further note of progress is the presence here of the wooden lockers, which came into use in army posts during the late 1880s.

[Negative Number RG-92-F-33-15, National Archives, Washington, D.C.]

Figure 3.19. Barracks, 10th U.S. Infantry, Company B or D, Fort Marcy, New Mexico, 1891-1895

In the 1890s, the troops at Fort Marcy, New Mexico, lived in a single-story barracks. The bunks in the sleeping area all appear to be the Composite Iron Works Company's Bunk No. 10, the one with the shorter footboard. Between the first and the second bunks on the right in Figure 3.19, we see an army-issue barracks chair, Model 1883. And we notice that the man who slept in the second bunk was the only one in this barracks who put a dust cover on his wall storage shelf. We also notice that his bunk is the only one in the barracks with a large black-and-white dog lying at ease on top of it, so he may have needed the dust cover.

The neatly kept wall shelf at the far end of the room has an attractive print applied as decorative edging. Numerous pictures hang on the walls, an individualistic treatment typical of civilian interiors of this period, but less common in military barracks.

The businesslike stove (left foreground) is the army's cast-iron wood heater No. 3, which first appears in specifications dated 1875. The footlockers are regulation, type 1875.

[Negative Number RG-F-36-8, National Archives, Washington, D.C.]

Figure 3.18

Figure 3.19

Figure 3.20

Figure 3.21

Figure 3.20. Cavalry Barracks, Post unknown

The wall shelves of the cavalry barracks shown in Figure 3.20 are typical of storage arrangements in army barracks of the late 1800s. Of particular interest here is the close-up view theses shelves provide of the kind of equipment and clothing the average man kept and the ways in which these things could be stowed away when not in use. In addition to the wall shelves, there were the footlockers, the army's standard Model 1875.

The bunks in this room are the Composite Iron Works Company's Bunk No. 10, dating from the 1880s (see Figure 3.15). Hanging from the ceiling at upper left and at the far end of the room are the two barracks lamps, the army's new Model 1891 two-burner. Strategically placed down on the open center aisle are the troopers' "spit boxes."

The three stoves in this long barracks room—the one best seen in the center distance and its two partners at opposite ends of the room—are the army's cast-iron coal heater No. VI.

Although we have no photograph of the outside of this barracks, identification data indicate that the building was a single-story adobe structure, typical of those built in the Southwest. Two such structures survive today at Fort Davis National Historic Site, Texas.

[Smithsonian Institution, Photograph Collection, Washington, D.C.]

Figure 3.21. Barracks, Fort Mason, California, 1891-1895

Here, again, as in Figure 3.18, we see, all at the same time, three types of nineteenth-century army bunks present in one barracks: the Composite Iron Works Company's Bunk No. 9 and No. 10, and the Snead Barracks Bunk, all with woven-wire springs. In this barracks, the order of the day was clearly to make mattresses and bedclothes into a bedroll to be stowed carefully at the head of the bed. (In this photograph, the photographer had removed the blankets from the bunk nearest the window and used them to control the light in the room as the film was exposed.)

Fort Mason's men had wall storage units with double shelves and (at right center), a surprisingly decorative iron stove (non-regulation). The footlockers, too, were non-regulation, although they are similar to the army's Model 1875 footlockers, only taller. The room's single hanging lamp is the army's Model 1891 two-burner pendant.

[Negative Number RG-92-F-37-5, National Archives, Washington, D.C.]

Figure 3.22. Cavalry Barracks, Fort McIntosh, Texas, 1891-1895

Figure 3.22 is the first photograph in this book to show a U.S. Army barracks equipped for lighting by electricity. At upper left, midway between the stovepipe and the fabric hanging above the gun rack, a clear, unshaded (and unlighted) incandescent electric light bulb is visible at the end of a long cord suspended from the ceiling. The reflection of very bright light from the rear wall of the barracks indicates other electrical outlets for lights there.

This photograph, dating from the 1890s, indicates that the army—or at least that part of it at Fort McIntosh, at that time—was fairly well in step with the rest of the country in installing electric lighting structures where peopled lived and worked. That could not be done until a practical incandescent electric light bulb and a practical system for providing it with electric power could be developed.

The first incandescent lamp intended for use in lighting homes electrically was demonstrated in 1859, at Salem, Massachusetts, by Professor Moses, G. Farmer. The experiment worked, but the product was not practical for manufacture or extensive use. Thomas A. Edison found the answer to lighting small indoor spaces electrically in 1879, when he produced the first successful incandescent light bulb. He built the entire range of equipment needed to operate and control it, and in 1882, opened the world's first central electric light power plant—the Pearl Street Station, in New York City. By the 1890s, electricity had begun to displace gas lighting as a way to illuminate homes, office buildings, small shops hotels, public structures, and military buildings, as well as streets.[17]

The fabric hanging above the gun rack in Figure 3.22 is something of a "first" as well. It is a dust cover, designed to keep the weapons of the soldiers in this barracks more nearly ready for instantaneous inspection. Its installers made the dust cover easy to raise and lower quickly. We see it raised here to illustrate its use.

The wood stove shown here in active use can be seen close-up in Figure 3.23. No specifications are available for this stove, but the model photographed in Figure 3.23 is an original, now kept at Fort Sam Houston Museum, Texas.

[Negative Number RG-92-F-39-4, National Archives, Washington, D.C.]

Figure 3.23. Military Stove, Fort Sam Houston, Texas

Figure 3.23 shows a present-day close-up of the iron body of the army's wood-burning "U.S." stove, seen in Figure 3.22 in active use at Fort McIntosh in the 1890s. The model shown here, an original, has lost its legs and the stovepipe fittings that connected it to a chimney for venting when it was in use.

The design for the large "U.S." plates on each side of the stove—and on each end, as well—may have been copied from insignia worn by U.S. Army enlisted men from 1814 to 1832, although it is doubtful that the stove is that old. It is a most interesting relic, and I hope that further research will turn up an exact date for it.

[Fort Sam Houston Museum, Texas]

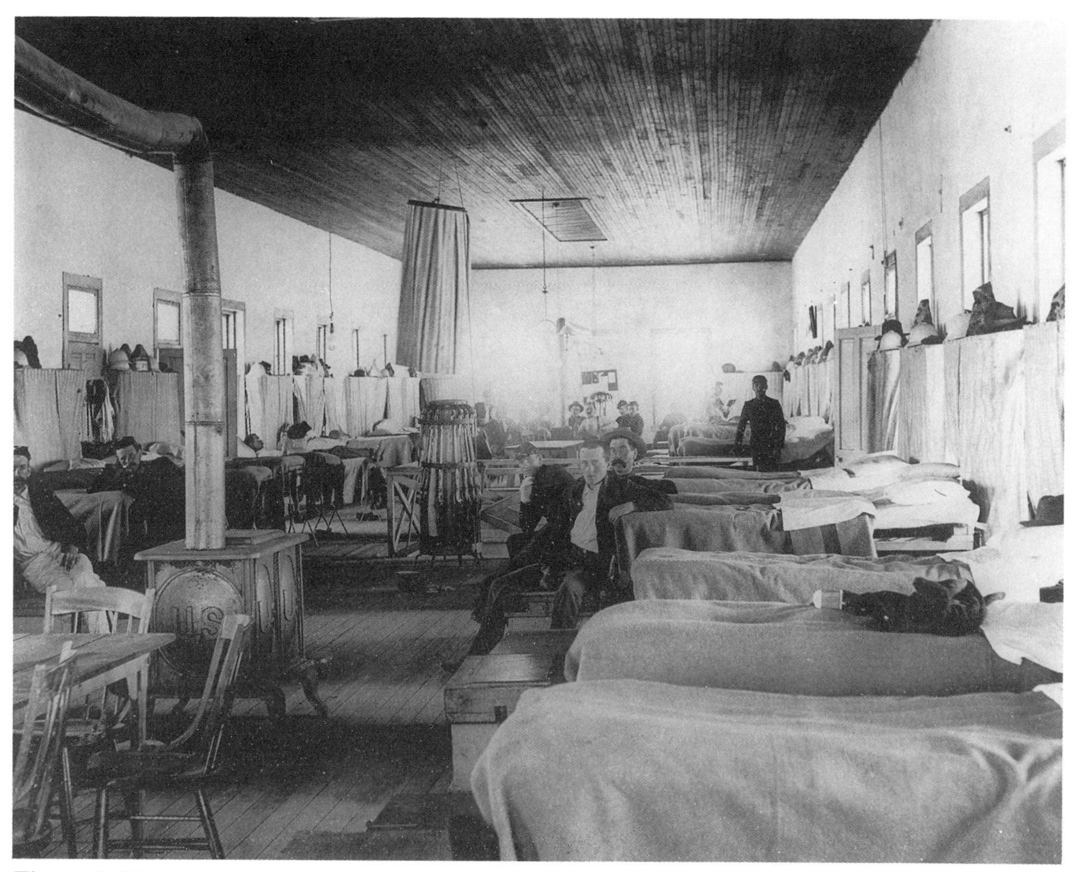

Figure 3.22

Figure 3.23

Figure 3.24

Figure 3.25

Figure 3.24. Cavalry Barracks, Fort Yellowstone, Wyoming, 1891-1895

What could be called a typical nineteenth-century army barracks room scrubbed, shining, and ready for inspection appears in Figure 3.24. The identical dust covers on all the wall shelves are probably made of the army's blue kersey or bunting. As noted earlier, the army's use of blue fabric as a protective covering for tables and desks and as draperies or curtains for windows was quite common. One officer remarked that this practice dated all the way back to the days of George Washington—and possibly it did. Blue was popularly held to be the "national color." The dyes used to produce popular shades of blue—including "army blue"—were usually durable, slow to fade, and readily applicable to the several kinds of long-wearing fabrics then available. A clerk's stand-up military desk, dating from around 1840 (now in the collection at Fort Scott National Historic Site, Kansas), still retains, clinging to its moldings, pieces of the original blue wool that covered it when it was issued.

Almost all the footlockers in Figure 3.24 are of the 1875 pattern. The rectangular stove is the army's cast-iron wood heater No. 2. The non-regulation spittoons, looking like deep dishpans, centered on the floor before and behind the stove, are made of treated fiberboard. The barracks lamp at upper center is an army Model 1891.

Wall clocks appear in many of the 1890s barracks photographs. The army's activities were conducted on a tightly coordinated time schedule, and time-keeping devices were both necessary and proudly displayed. Before inexpensive, reliable clocks were available, the beginning and ending times for various required activities throughout the day were signaled by a company fifer, drummer, or bugler who announced them by playing the officially designated musical or rhythmical calls—a procedure followed well into the twentieth century.

[Negative Number RG-92-85-4, National Archives, Washington, D.C.]

Figure 3.25. New Barracks (Cavalry), Fort Custer, Montana, 1891-1895

The photograph of Fort Custer's cavalry barracks (Figure 3.25), new in the early 1890s, provides a good, clear view of the barracks wall lockers that were gradually superseding the long-used open shelves attached to barracks walls for the soldiers' gear and spare clothing. These then-new wall lockers can also be seen in Figure 3.18, in use at the Fort Leavenworth, Kansas barracks during that same period. Fort Custer's wall lockers appear to be, basically, neatly boxed-in wall shelves with two hinged doors on the front and a downward slanting top added to each unit, all attractively finished with two coordinating colors of paint. Whether their tops were made to slant forward in the interest of neatness—nothing carelessly tossed up there could cling there very long—or whether that oddly slanting top was made for aesthetic reasons, only their designer knows.

The bunks here are mainly the Composite Iron Works Company's Bunk No. 9. All footlockers appear to be the army's Model 1875. The two big iron stoves are non-regulation.

On the right door, facing the big, open entry way (left foreground), is the first barracks bulletin board seen in this collection. Clearly well-used, it is covered with notices. On the back wall are two framed rectangular racks made to hold training whips used by the troopers in working with horses. No other equipment used in training or riding horses appears in this cavalry barracks. Saddles, bridles, and other gear used with the horses would probably have been kept in the tack room at the fort's stables.

One might search the walls very closely here and still not find the barracks-room clock, but the "facts" file for this photograph states that the clock is the framed object jutting from the right-hand wall between the first and second window frames. The file notes that this clock appears to be an eight-day Ogee style.

[Negative Number RG-92-F-17-14, National Archives, Washington, D.C.]

Figure 3.26. Infantry Barracks, Fort Custer, Montana, 1891-1895

Black infantrymen, off duty, relax in their single-story barracks (Figure 3.26) at Fort Custer in the 1880s. They were soldiers in either the 24th or the 25th Infantry; two black regiments organized and put into service by the army in 1866. Both of these regiments, as well as the two black cavalry regiments activated at the same time—the 9th and 10th Cavalry—served meritoriously throughout the thirty years of the Indian Wars in the trans-Mississippi West (see "The Buffalo Soldiers").

Although both black and white soldiers on frontier duty in the late 1800s sometimes drew hardship posts, the army's Quartermaster Corps made every effort to provide equipment and furnishings for them identical to those of all other enlisted men's barracks at all comparable army posts. Here, two-toned wall lockers with hinged doors provide storage for clothing and personal possessions. The bunks are Composite Iron Works Company Bunks No. 9; the barracks lamp is the army's new Model 1891; the footlockers are army model 1875; the two big stoves are non-regulation. A well-utilized bulletin board—difficult to see—is attached to the left door-facing the entry way (left foreground). Roller window shades screen the two windows on the left wall. Only the spittoons in this photograph are a bit out of the ordinary. They appear to be a new kind, made of metal, with the outer surface covered in porcelain, and they had just come into inventory in 1893.

[Negative Number RG-92-F-17-15, National Archives, Washington, D.C.]

The Buffalo Soldiers

During the Revolutionary War (1775-1783), many black men fought in both the Continental army and navy. Beginning in 1862, black men could—and did—enlist to fight in the Civil War in the U.S. Regular Army. Throughout the Civil War, more than 186,000 of them fought in the Union ranks. In 1866, four black regiments—the 9th and 10th Cavalry and the 24th and 25th Infantry—were organized and immediately put into action by the army. For the next thirty years, the black soldiers saw continuous service all over the West, the Southwest, and the Great Plains, during the last decades of the Indian Wars.[18]

The Indians, an eloquent people whose gift for apt descriptions made all white men "palefaces," all army men "Bluecoats," and all cavalry troops "Pony Soldiers"—immediately dubbed the black men the "Buffalo Soldiers" because the black men's short, tightly curled dark hair, seemed, to the Indians, like that of the buffalo. When they heard about the nickname, the men of the 10th Cavalry promptly adopted the buffalo as the central figure in their regimental crest and served under it proudly.[19]

Many nineteenth-century white soldiers regarded army service as temporary, make-shift work, but the black men welcomed it as one of the few career choices hey had at that time, and went about it with courage, determination, endurance, and buoyant spirits. Although they were often assigned "hardship posts," few black men deserted the service, and their reenlistment rates were consistently high.[20]

U.S. armed forces remained racially segregated throughout World War II. Progress toward integration in the services, begun during President Franklin D. Roosevelt's administration (1933-1945), was continued under that of President Harry S. Truman (1945-1953). Truman's 1948 directive calling for an end to segregation in the U.S. armed forces did much to accomplish that, and it was finally achieved completely in 1953.[21]

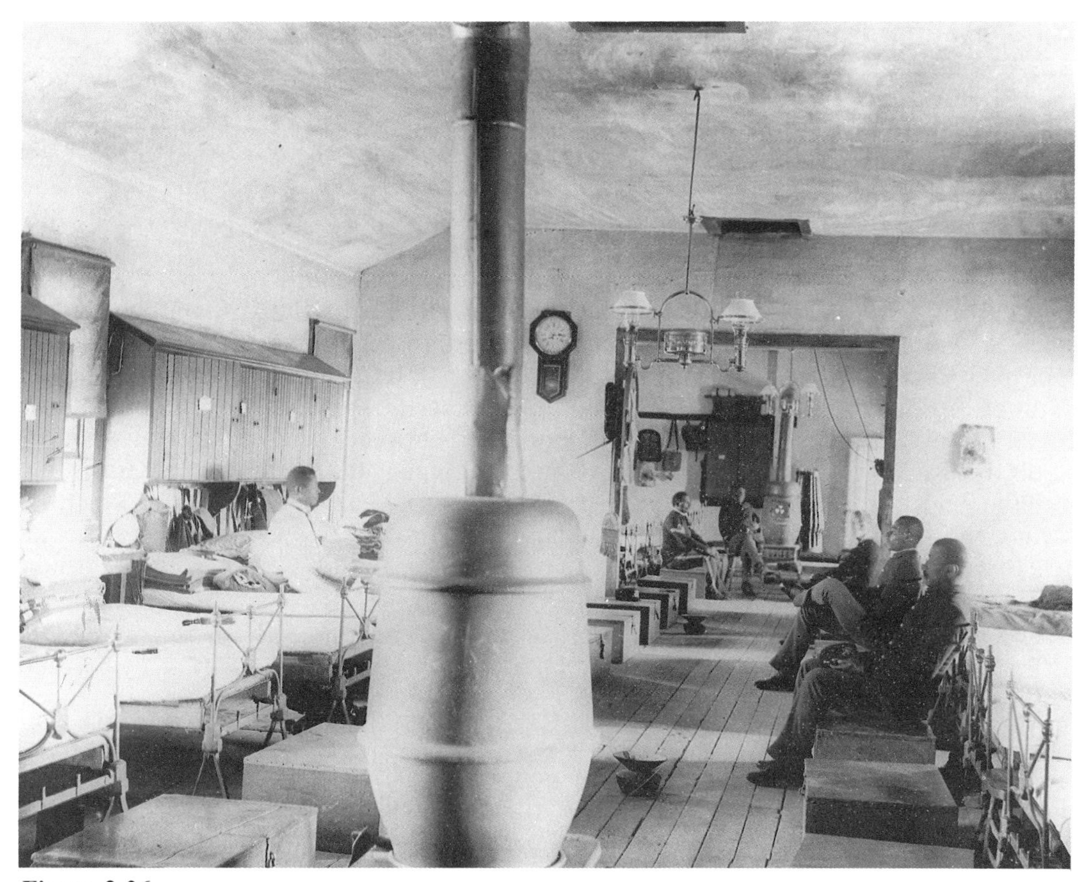

Figure 3.26

Figure 3.27

Figures 3.27 and 3.28. Fort Huachuca, Arizona, 1891-1895: Cavalry Barracks and Infantry Barracks

Isolated army posts, particularly those along the far-reaching frontier of the west and the great plains, often needed the combined expertise, range, strength, and speed of movement of both infantry and cavalry, hence, garrisons were often manned by both mounted troopers and foot soldiers. Sometimes both infantry and cavalry were of the same race; sometimes, as was true at Fort Huachuca and Fort Custer, one unit would be black and the other white. Figure 3.27 shows the Fort Huachuca cavalry barracks (white) and Figure 3.28, the infantry barracks (black). Some garrisons also maintained a full unit of Indians who had enlisted as soldiers in the regular army (see Figure 3.29). Racial segregation was a fact of life in the U.S. armed forces until well after World War II, and sometimes, at some frontier posts, in addition to the routine rivalry between men wearing the same uniforms but practicing different specialties, there was also racial tension.Even with the racial segregation maintained in the 1800s, however, the army made an attempt to provide equipment and furnishings that were identical throughout for all military quarters for the men in uniform. The bunks in each of the Fort Huachuca barracks buildings appear to be Composite Iron Works Company's Bunk No. 10; the stoves are the army's wrought-iron Wood Heater No. 5 (an original of this model survives today at Fort Laramie National Historic Site, Wyoming).

Most of the footlockers are the army's Model 1875. The lamps in both barracks are the army's newer Model 1891. The wall clocks in both appear identical. In the infantry barracks, a wall chart about uniforms is just discernible beneath the clock on the back wall, and the infantrymen also have (almost excluded by the camera) a bulletin board at the head of the stairway.

[Negative Numbers RG-92-F-27-9 and RG-92-F-27-10, National Archives, Washington, D.C.]

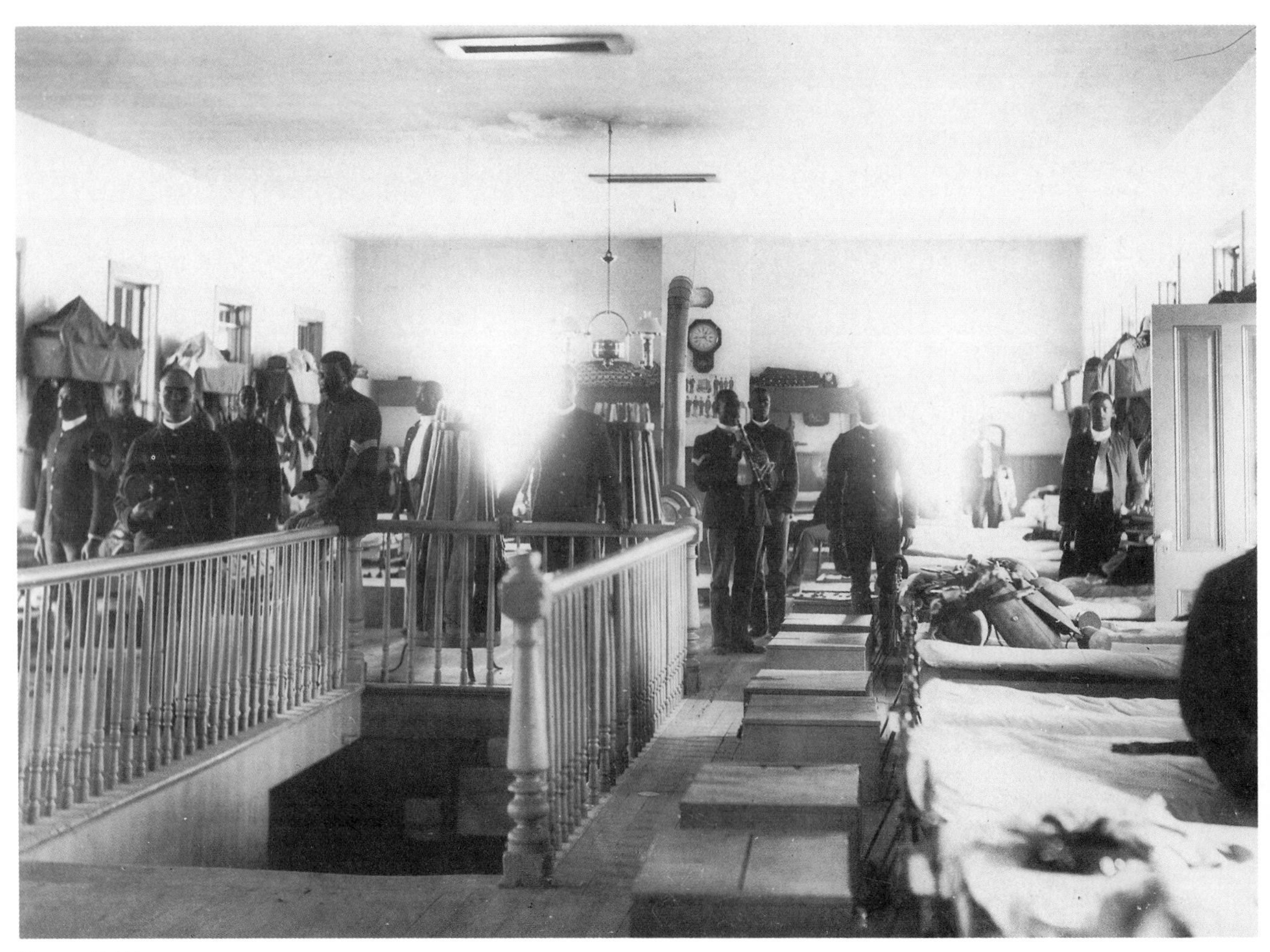

Figure 3.28

Figure 3.29. Indian Soldiers' Barracks, Fort Douglas, Utah, 1891-1895

The photograph in Figure 3.29, made at Fort Douglas in the early 1890s, shows a view comparatively rare in nineteenth-century military housing; the barracks of Indian enlisted men serving full time as U.S. Army Regulars. Indian civilians in great numbers had been hiring on for temporary work with the army as scouts and interpreters or as extra combat troops in random campaigns since the early 1800s; but not until near the end of the nineteenth century, after decades of war between U.S. troops and Indian warriors had finally ended, were there many Indian soldiers in the U.S. Army. At about the same time—the 1880s and 1890s—young warriors also began joining the reservation police force, then being established, to enforce reservation laws for both Indians and whites.[22]

The nineteenth-century army in the trans-Mississippi West routinely utilized the recognized skills of both civilian and military Indian allies as scouts, guides, and interpreters and as formidable fighting men when the need arose (see "Warriors in Uniform").

Although U.S. armed forces were racially segregated from early times until 1953, the Army Quartermaster Corps consistently made every effort to provide all soldiers with identical equipment and furnishings, regardless of the men's race or color.

In Figure 3.29, the two fat, rounded stoves warming the Indian soldiers' barracks (right foreground and center rear) are the army's cast-iron Heater No. 6, seen earlier in photographs of other barrack interiors. The room's two lamps are the army's newer Model 1891 two-burner barracks lamps. The circular arms racks near the two stoves date from about 1880. The footlockers are army standard, as are the wall shelves, here neatly and uniformly screened with cloth dust covers. The bunks are the army's standard Snead Barracks Bunks, covered here with the army's standard rubber blankets or ponchos.

In fact, the only item in this photograph that is not standard government issue is the company's mascot—the handsome cat regally surveying the scene from a seat on the table at center, just beyond the nearer gun rack.

[Negative Number RG-92-F-20-5, National Archives, Washington, D.C.]

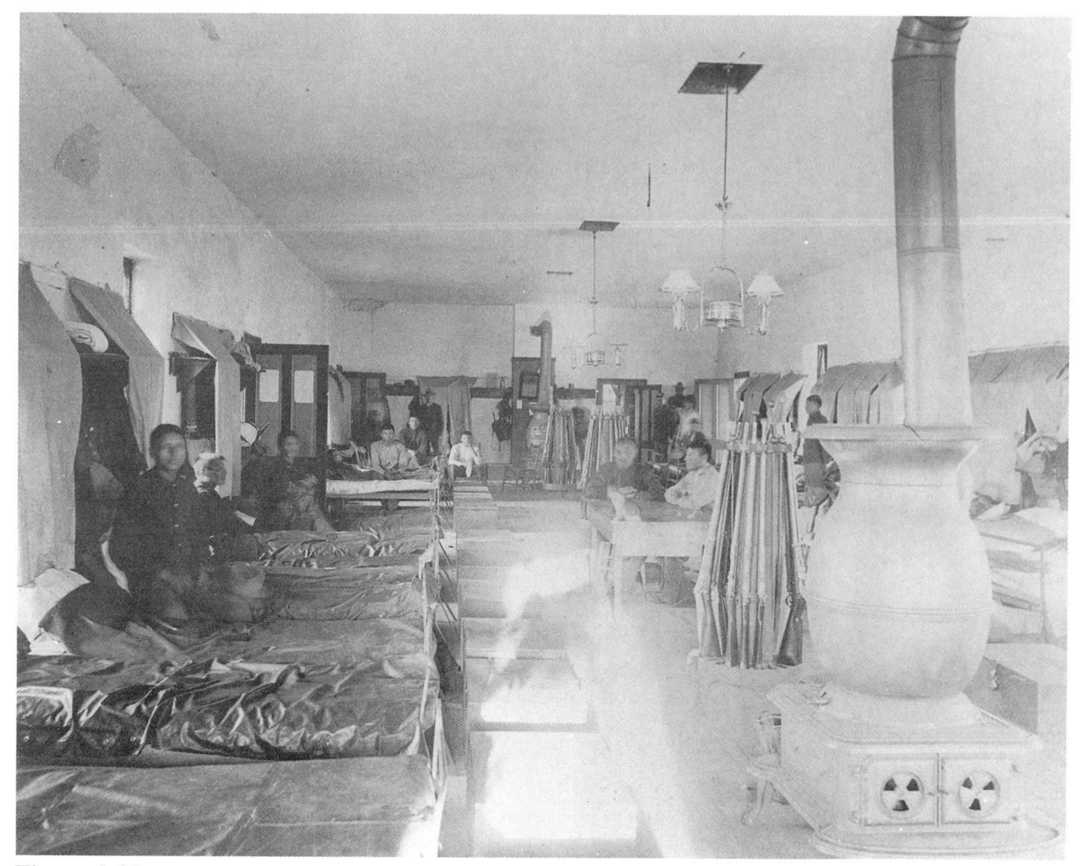

Figure 3.29

Warriors In Uniform

Few photographs of nineteenth-century Indian soldiers' army barracks, such as the one in Figure 3.29, are known to exist—probably because until near the century's end, Indian enlistment for full-time service in the U.S Regular Army had been negligible. Although Indian civilians in great numbers worked, throughout the 1800s, as paid employees of the army on temporary assignment as scouts, guides, interpreters, and occasional campaign combat troops, most of them did not give up their civilian status to become full-time, uniformed servicemen.[23]

In the latter half of the nineteenth century, however, the last traditional bastions of North American Indian life underwent drastic change. By the 1880s, traditionally limitless Indian hunting grounds had been severely restricted or taken for use by white settlers, land speculators, road builders, cattlemen, miners, and townsmen steadily pushing westward. Dealers in hides and suppliers of meat for the transcontinental railroad section gangs had destroyed the buffalo, the Indian's primary source of food, shelter, tools, and clothing. With tribal ranks critically depleted by warfare, disease, starvation, and exposure during the punishing Great Plains winters, more and more tribes were settling into reservations run by the U.S. government.[24]

Many young warriors there joined the reservation Indian police, established by the reservation agents of the Bureau of Indian Affairs to enforce the law for both the Indians and whites on and around the reservations. Others enlisted in the U.S. army.[25]

Whether in uniform or in native dress, Indian scouts worked with the soldiers on reconnaissance, exploratory marches, territorial or transportation route surveys, escort duty for supply trains, parleys with tribal chiefs, and armed expeditions against occasional insurrections by renegades or hostile tribesmen. Accustomed to continuing warfare between traditional tribal enemies—even to steady conflict within a single tribe—both civilian and military Indian allies gave full loyalty to their work with the army against old enemies.[26]

The Pawnees, long-time foes of the Sioux, the Cheyenne, and the Arapaho, scouted regularly for the army throughout the last thirty years of the Indian wars. Among the fiercest of fighters, the Pawnees never warred against the United States, even when the government treated them unjustly. During the 1860s, when the Union Pacific Railroad was under construction, the Pawnee scouts, including an elite group called "the Pawnee Battalion," protected railroad construction crews from raids by other Indians.[27]

During the Civil War, Utes scouted for Kit Carson, then a colonel in the Union Army, assigned to control hostile and renegade Comanches and Navajos in Arizona, Texas, and New Mexico.[28]

Throughout the 1870s and 1880s, General George Crook, notable for great patience and integrity in dealing with the Indians, relied heavily on his Apache scouts against renegade Apaches as well as against their old enemies, the Sioux. Sometimes nearly half of General Crook's fighting force was made up of his trusted scouts, with many Utes, Crows, and Shoshones included among the Apaches. When Crook's superior officer reprimanded him stiffly about relying more on his Indian scouts than on regular army units available, Crook resigned. His successor, General Nelson A. Miles, also relied on Indian scouts, fielding some five hundred Apaches with his regular civilian militia and soldiers on one expedition after Geronimo.[29]

And the Crows, who scouted for Colonel George A. Custer's 7th Cavalry gave the colonel full and accurate reports about the strength of the Sioux and Cheyenne forces gathered on the banks of little Bighorn in June of 1876.[30]

During World War II, Native Americans once again served U.S. Military forces willingly and well. The National Geographic Society's *World of the American Indian* observes:

> Pearl Harbor sparked the ancestral warrior spirit, calling forth a flurry of Indian Volunteers. Eventually some 25,000 served. A Creek and a Cherokee, Ernest Childers and Jack Montgomery, of the famed Thunderbird Division, won the Congressional Medal of Honor. ...General Clarence Tinker of the Army Air Corps, an Osage, died in the Pacific. Ira Hays, a Pima Marine, helped raise the flag at Iwo Jima. ...A Pawnee, Brummet Echohawk, a renowned expert in hand-to-hand combat, trained commandos...[and] Navajos...encoded messages in their native tongue.[31]

Figures 3.30, 3.31, and 3.32. Arms Racks in Infantry Barracks Buildings, 1891-1895: Key West, Florida; Fort McPherson, Georgia; and Fort Snelling, Minnesota

Figure 3.30, 3.31, and 3.32 are grouped together to show the different types of gun racks used in army barracks buildings dating from the same period. The racks in Figures 3.30 and 3.31 are the kind often found in older barracks buildings, dating before 1870. Figure 3.32 has a beautifully clear view of the army's Model 1875 rack, as well as another item appearing in this collection for the first time: the strange looking object perched on a pipe, between the handsome gun rack and the stove, is a hot-water heater. Through the pipe on which it stands, it makes hot water available in the soldiers' washroom directly below. (A closer view of this kind of water heater appears in Figure 3.75.)

And a closer look at the Model 1875 gun rack in Figure 3.32 reveals identification numbers stenciled on the outside rim of the shelf at the bottom of the rack. The same number that appears under each stacked rifle would also have been stenciled on the clothing and other equipment of the soldier to whom the gun was issued.

In Figure 3.30, close inspection will reveal that the woven-wire bedsprings are not all secured in the same place on the No. 10 Composite bunks, and the soldiers in Figure 3.30 appear to be using rubber ponchos as dust covers for their wall storage shelves.

[Negative Numbers RG-92-F-32-4, RG-92-F-41-20, and RG-92-F-65-7, National Archives, Washington, D.C.]

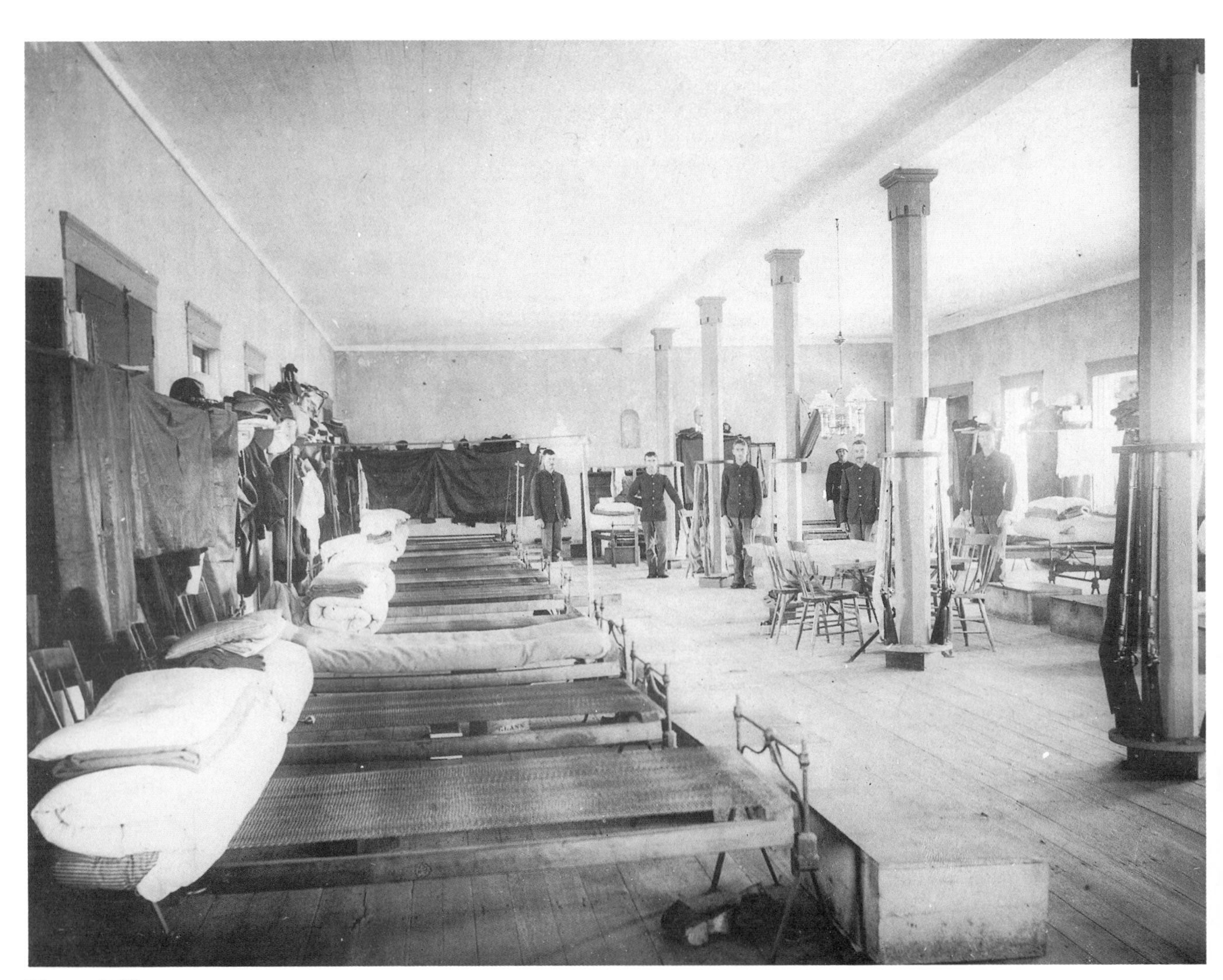

Figure 3.30

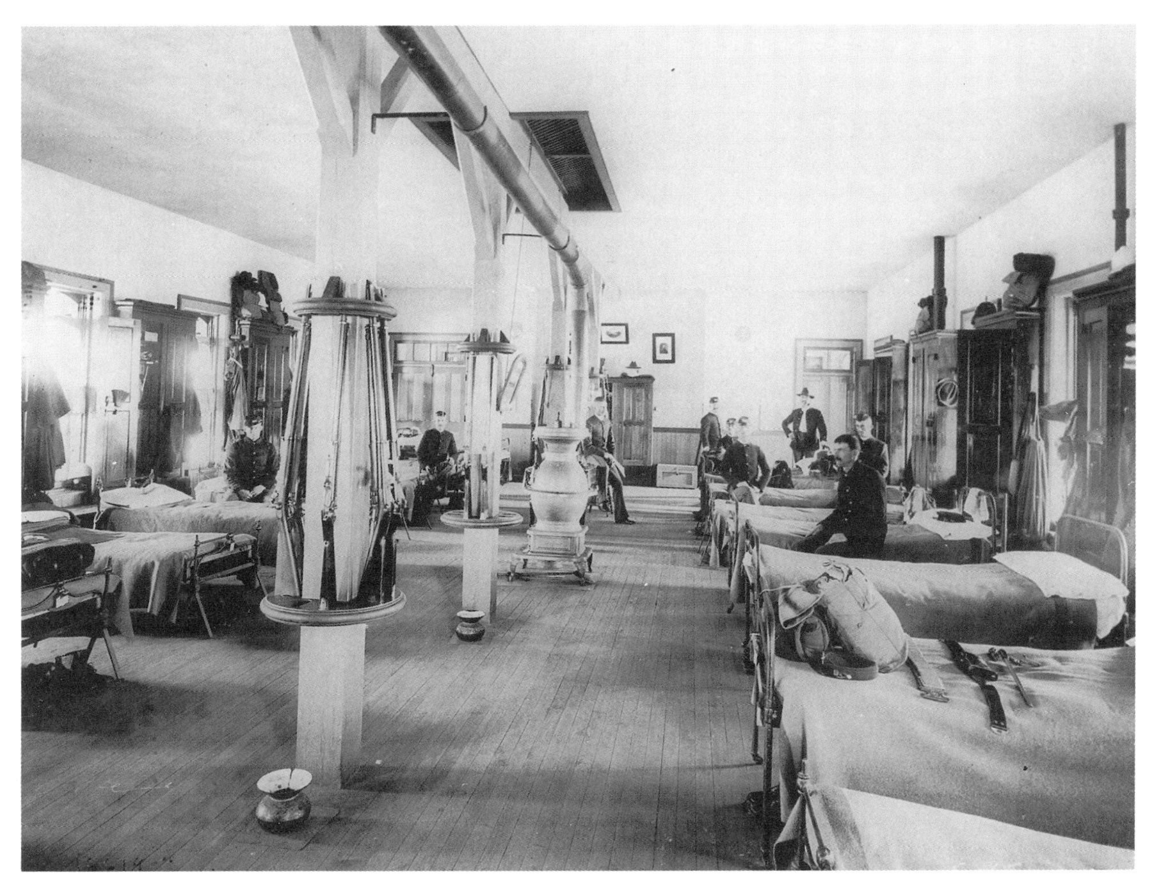

Figure 3.31

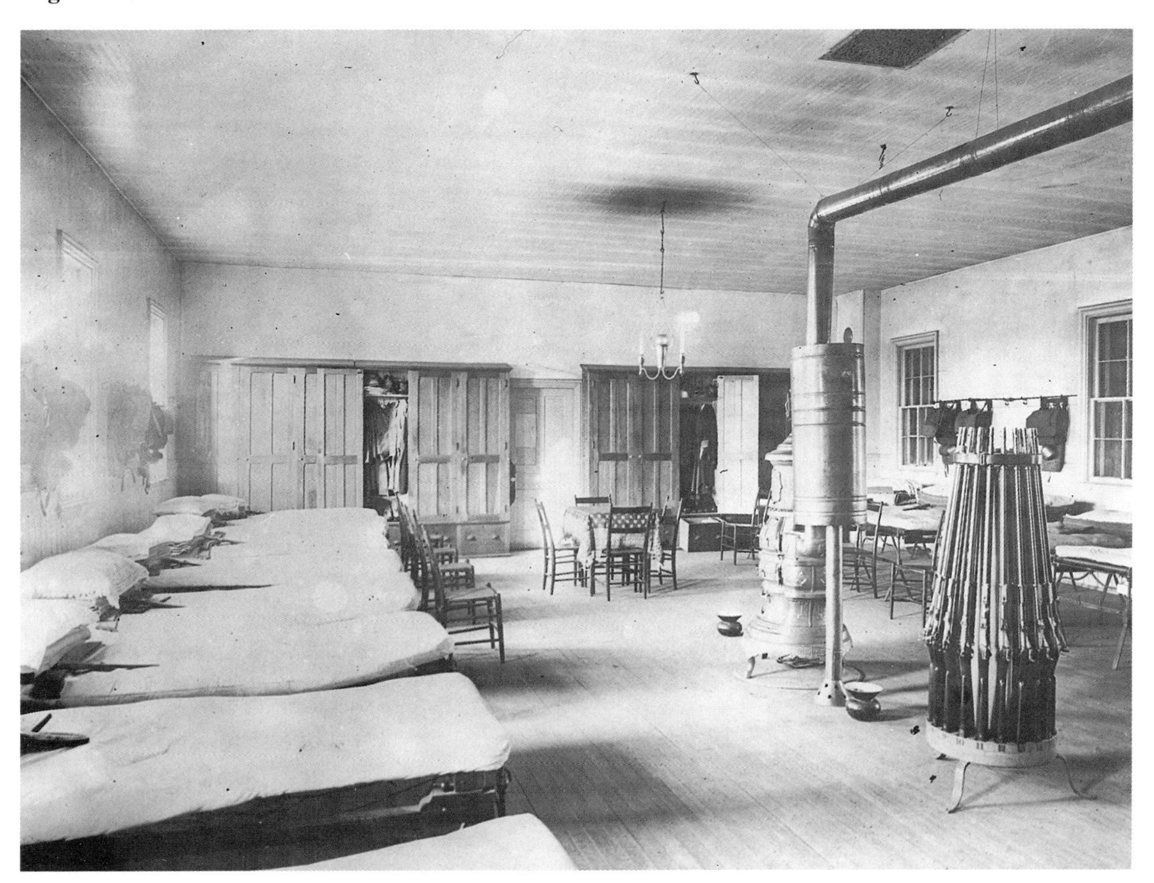

Figure 3.32

Figure 3.33

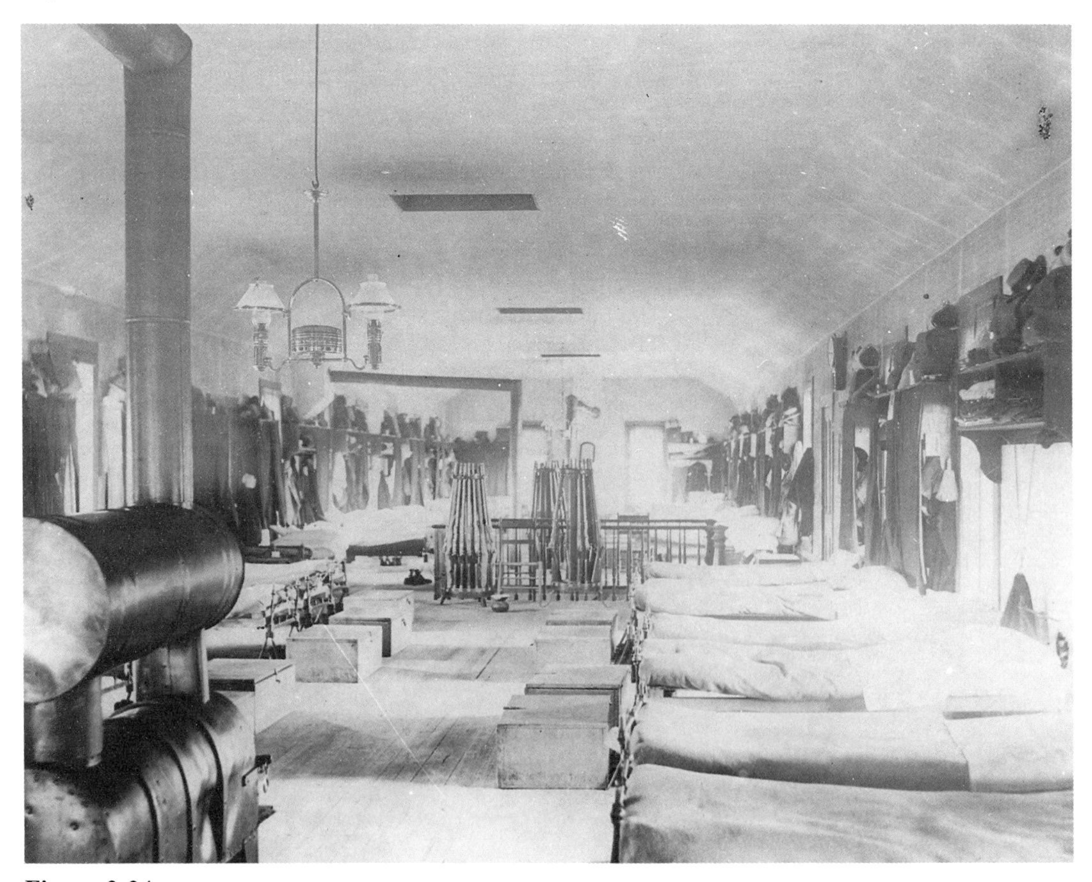

Figure 3.34

Figure 3.33. Infantry Barracks, Fort Assiniboine, Montana, 1891-1895

The enormous stove that reaches toward the ceiling and almost upstages the gun rack (right foreground) in Figure 3.33 is the army's Model 1875 cast-iron coal heater, either No. 6 or No. 7.

Storage space for the men's extra clothing and personal belongings here seems to have been limited to wall pegs and footlockers, although it is possible that wall shelves or wall lockers were provided in an area that the camera did not include.

Noting the smoke shields—circular metal attachments that look like percussionists' cymbals—suspended above the Model 1891 barracks lamp (rear center), one also notices, immediately, the chromolith of Napoleon Bonaparte, hanging high on the rear wall, a surprisingly witty and appropriate barracks room decoration.

The company bulletin board on the rear wall, to the right of the doorway, is just barely visible, almost blocked out by the form of the company bugler, sprawling at ease on the table top in front of it. And the barracks clock, which only the eagle-eyed may detect on the shelves atop the desk at the rear right, appears to be a Gothic steeple design.

[Negative Number RG-95-F-5-5, National Archives, Washington, D.C.]

Figure 3.34. Infantry Barracks, Vancouver, Washington, 1891-1895

Two odd items appear in Figure 3.34, one on the left foreground, the other in the background at far right. The one in the left foreground is the army's wrought-iron wood heater No. 4, Model 1975, with a metal drum attached to its top by pipes. The other, in the far right background, is a civilian straw hat—barely visible because of the camera's distance from it—on the top shelf.

The metal drum on the top of the heater is a ''heat Chamber,'' and was put there in an effort to boost the stove's efficiency at producing heat. The idea was that the heat chamber would catch the hot smoke rising upward from the fire in the stove and detain it long enough for the smoke to contribute a few degrees more warmth to the room as it circulated in the chamber before escaping up the chimney.

The civilian straw hat, a modish style of relaxation among the uniform caps in this infantry barracks, was of a type much in vogue with men of the 1890s, both soldiers and civilians. Made of stiff, woven straw, it had a flat, level, inflexible brim; a flat, low crown; a wide, brightly colored ribbon band; and since it was the fashionable summer headgear for men of that era to wear on Sunday afternoon boating excursions with their best girls, it was frequently called a ''boater'' or a ''sailor hat,'' or in some areas a ''strawkady'' (or ''katy'').[32] It was extremely popular with the soldiers, who wore it even in uniform if they were off duty, and the season for it was as carefully observed as that governing the ladies' Easter bonnets and white shoes. It was worn everywhere throughout the summer, and on Labor Day afternoon, after the last ''out'' of the Labor Day baseball game, every boater in the baseball park was scaled out onto the diamond.[33]

[Negative Number RG-92-F-71-6, National Archives, Washington, D.C.]

Figure 3.35. Artillery Barracks, Fort Trumbull, Connecticut, 1891-1895

The army's cast-iron coal heater, Model 1875, No. 6 (or No. 7), can be seen from both the front and rear in Figure 3.35, with the rear view in the immediate foreground. The personnel at Fort Trumbull have put protective rails around the stove as a general safety measure.

We notice the army's newer, round, portable gun rack at center, and looking toward the far end of the room, we notice also that the round supporting pieces of the old gun racks have been left where they were originally attached on the last four roof-support posts.

The ship model suspended from the top of the room's third roof-support post is a fitting touch in this coastal fort.

[Negative Number RG-92-F-70-4, National Archives, Washington, D.C.]

Figure 3.36. Infantry Barracks, Fort Niagara, New York, 1891-1895

The bunks in the barracks photographed in Figure 3.36 are routine army issue; all of them are Composite Iron Works Company's No. 10 models. They appear, however, in a barracks whose inhabitants have a touch of individuality. A close look at the left wall reveals that these Fort Niagara soldiers are using half-sections of pup tents as dust covers for their wall shelves, the metal grommets in the tent fabric can be seen at the top and bottom of the two strips caught by the camera.

The potted plant flourishing at far left is a rare possession in a nineteenth-century army barracks, as are the nicely decorated roller shades on the three windows along the rear wall.

[Negative Number RG-92-F-46-7. National Archives, Washington, D.C.]

From Circular No. 4, *Report on Barracks and Hospitals.*

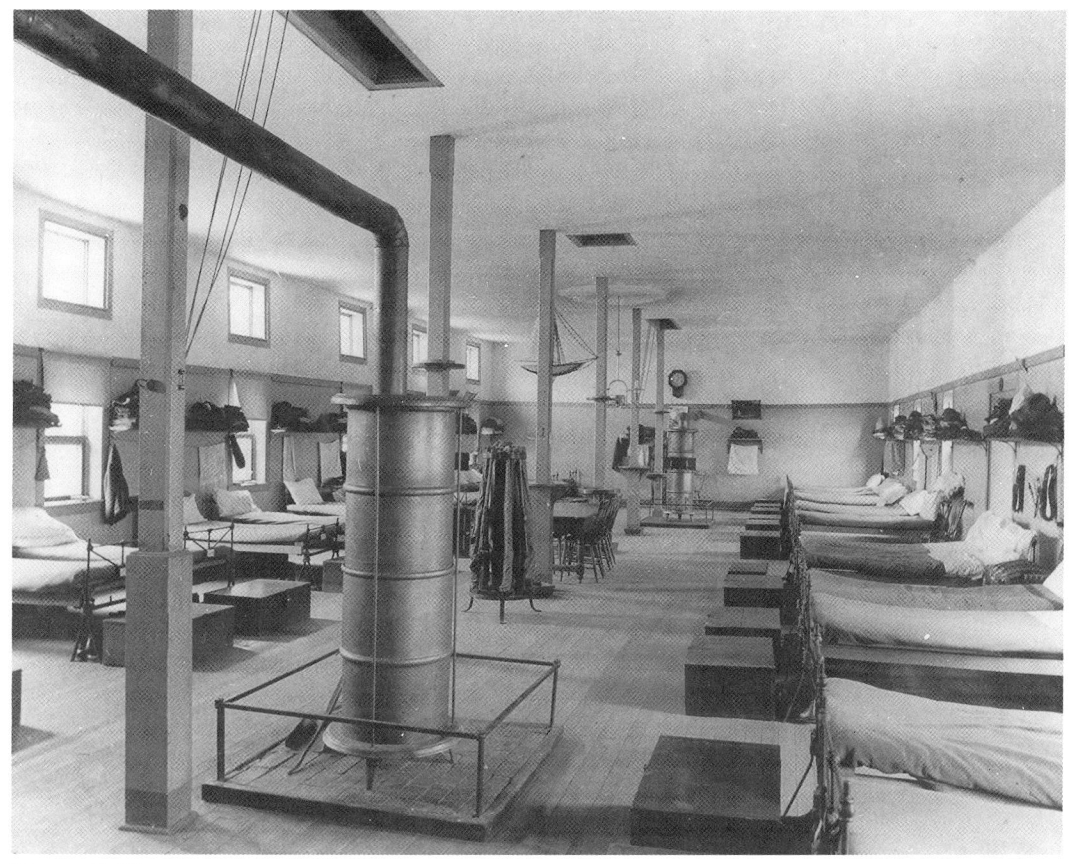
Figure 3.35

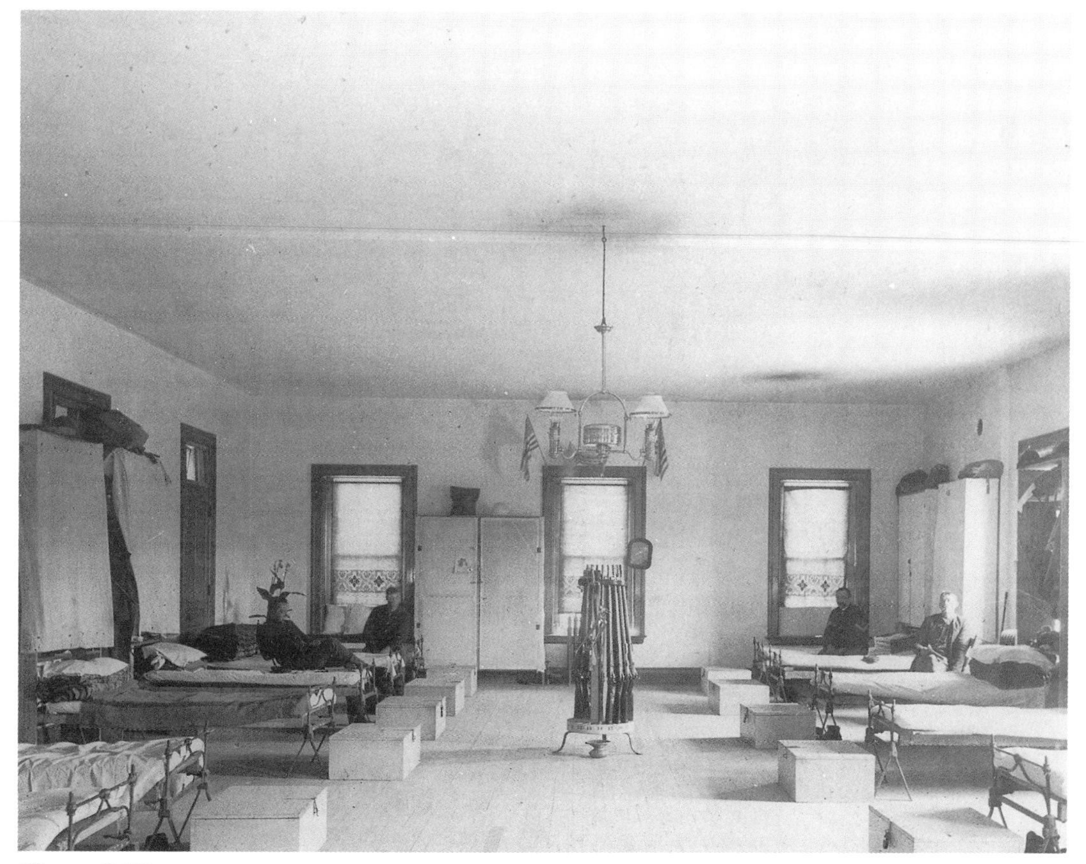
Figure 3.36

Figure 3.37

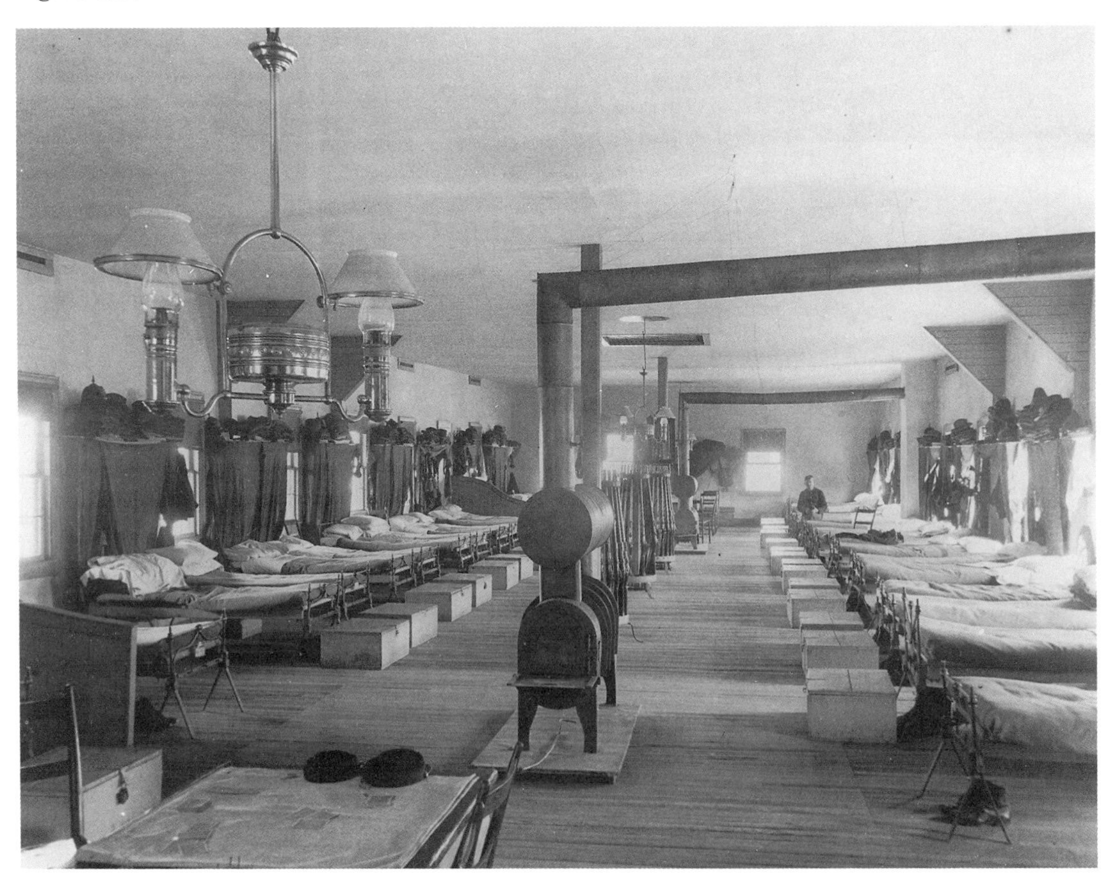
Figure 3.38

Figure 3.37. Cavalry Barracks, Fort Meade, South Dakota, 1891-1895

The beds and stoves in the barracks in Figure 3.37 are the items most out of the ordinary. The beds have half-shelter lengths of pup tent canvas use on them as bed covers, and the height at which the woven-wire bedsprings are secured to the framework of the No. 10 Bunks is unusual. It raises the mattresses and bedding above the top of the iron bunk frames, not an arrangement usually seen.

The two barracks stoves seen here are the army's cast-iron wood heater No. 2, both of them with heat chambers. These heat chambers, however, are different from the ones seen in Figure 3.34. The heat chambers in the earlier photographs appear to be solid cylindrical metal objects, much like large metal oil drums. The ones in this photograph look like fat automobile tires fitted between the stove and the stovepipe.

Both types were experimental efforts to obtain more warmth from the fires in the stoves by causing the rising smoke to circulate long enough on its way up the chimney for the thin metal walls of the heat chamber to absorb that warmth and later radiate it out into the atmosphere of the room. The heat chambers in Figure 3.37 may possibly have had a slight advantage over the earlier ones; these tire shaped types look as if they might have provided more air-flow space for the smoke to linger in than the drum types in the earlier photo.

[Negative Number RG-92-F-42-9, National Archives, Washington, D.C.]

Figure 3.38. Infantry Barracks, Fort Townsend, Washington, 1891-1895

The most distinctive feature of Figure 3.38 is the splendid close-up (upper left foreground) of the army's two-burner barracks lamp, Model 1891, the best shot yet seen of this handsome, well-designed light fixture.

Directly beneath the lamp, two enlisted men's caps and scattered playing cards, face down on an abandoned card table, indicate that two soldiers stopped their card game and left the table so that the photographer might have a clear view of the room and its furnishings. A folded canvas shelter-half is used as a table cover.

And, here again, as in Figures 3.34 and 3.35, we see army barracks stoves with heat chambers. The stoves in Figure 3.38 are the army's wrought-iron wood heater No. 5 (or No. 6), Model 1875, with the solid "oil drum" type of heat chamber, like those in Figure 3.34.

The footlockers in this barracks are all army standard, Model 1875.

[Negative Number RG-92-F-69-5, National Archives, Washington, D.C.]

Figure 3.39. Infantry Barracks, Fort Clark, Texas, 1891-1895

Even as late in the nineteenth century as this photograph was taken, some barracks were still using non-regulation stoves, such as the one here, at right foreground. And the tall wooden civilian wardrobe against the rear wall, beyond the second stove and the last gun rack at the far end of the room, is also a piece of furniture not usually seen in an army barracks.

A third point somewhat out of the ordinary about this barracks is the location of the footlockers; they are under the soldiers' bunks. They are there because for a barracks room, this is exceptionally narrow, and footlockers, in their usual place, at the foot of each bunk, would have made the center aisle passageway uncomfortably narrow for the number of men who needed to move through it.

[Negative Number RG-92-F-14-4, National Archives, Washington, D.C.]

Figure 3.40. Cavalry Barracks, Fort Walla Walla, Washington, 1891-1895

Figure 3.40 reveals Fort Walla Walla as another 1890s fort with a non-regulation barracks stove, non-regulation footlockers, and another instance of the odd high placement of the woven-wire bedsprings on their Composite No. 10 bunks.

The windows of this second-floor barracks are fitted with roller shades, most clearly visible on the second and third window of the right-hand wall, and the barracks lamp is the army's older Model 1881.

[Negative Number RG-92-F-73-5, National Archives, Washington, D.C.]

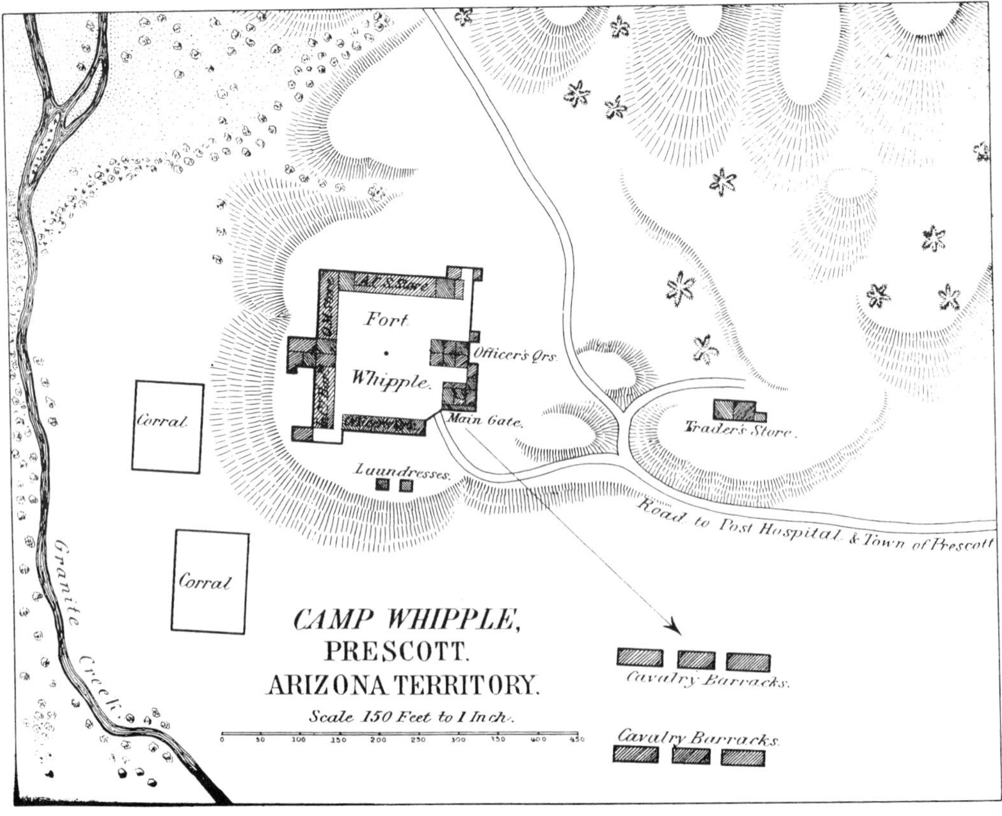

From Circular No. 4, *Report on Barracks and Hospitals.*

Figure 3.39

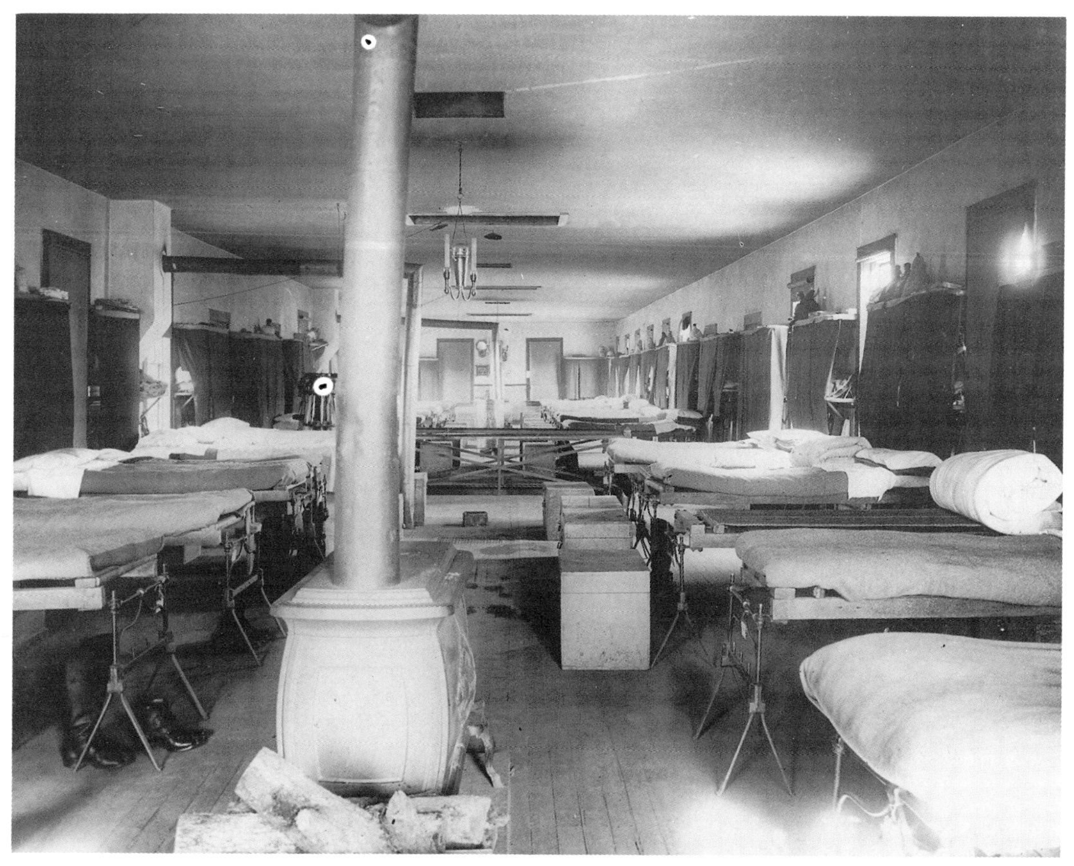

Figure 3.40

Figure 3.41

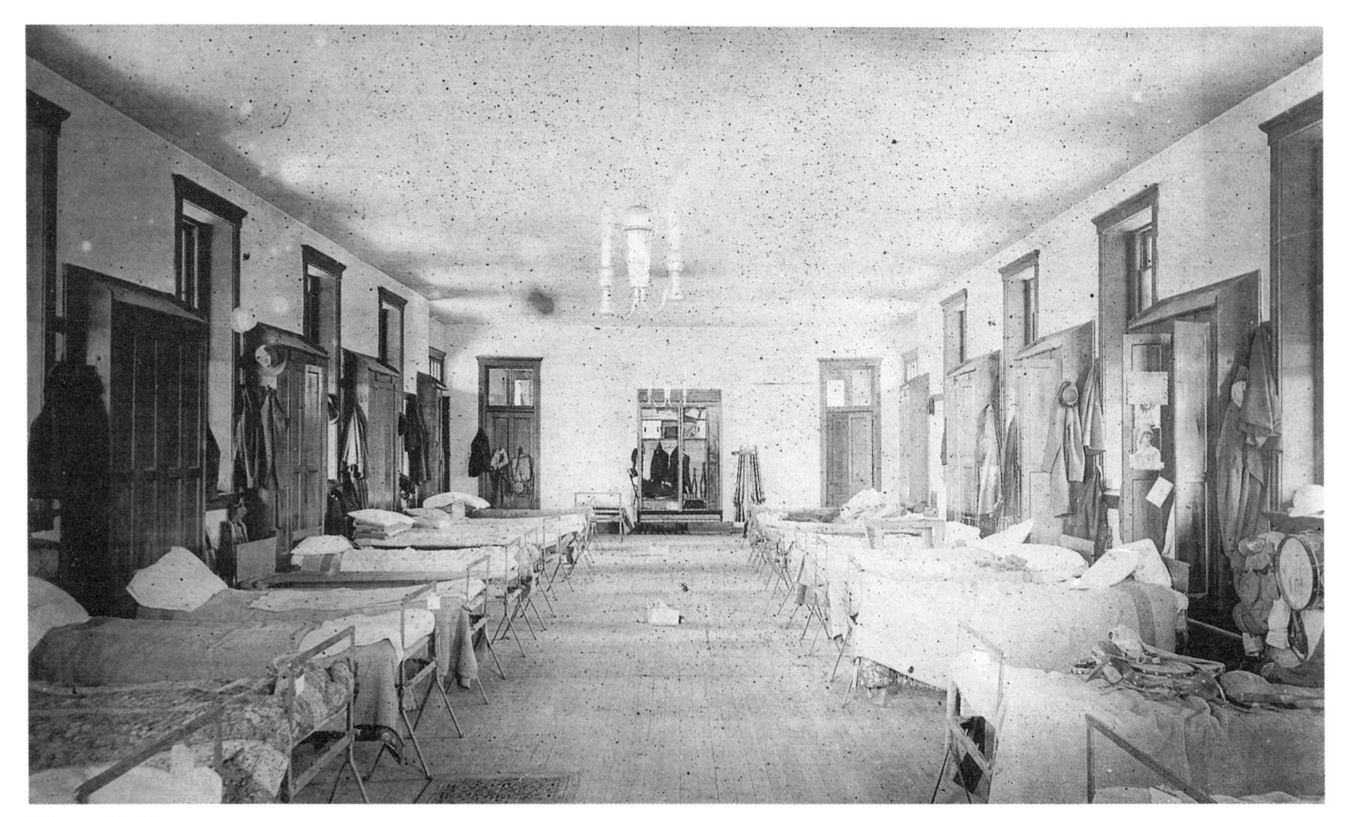

Figure 3.42

Figure 3.41. Whipple Barracks, Arizona (Infantry), 1891-1895

The stove in Figure 3.41 might be the army's regulation officers' parlor stove, Model 1875, though positive identification is not immediately available.

The room itself, with its high archway between the main dormitory space and the small, almost secluded, windowed area at rear, is anything but regulation.

The beds here are a mixed lot, partly the Composite Iron Works Company's No. 9 bunks and partly the Snead Barracks Bunks. The use of lengths of tent canvas—"shelter halves"—as bed covers and dust covers is quite evident; the metal grommets around the edges of the canvas are plainly visible. The other pieces of fabric used as dust covers in this room are interesting examples of the inexpensive textile prints available throughout the late 1800s.

[Negative Number RG-92-F-79A-5, National Archives, Washington, D.C.]

Figure 3.42. Infantry Barracks, Fort Logan, Colorado

Either Fort Logan's infantry barracks was photographed in the summertime in Figure 3.42, or it had furnace heat; no stoves of any sort, regulation or otherwise, are in evidence. The only gun rack to be seen is stationed out of the way against the back wall, and the long center aisle is free of all traffic impediments except for two "spit boxes," the second one very near the back and difficult to detect.

All the beds visible are the Snead Barracks Bunks. The room is lit by two older army barracks lamps, both Model 1881. The second lamp, like the second spittoon, is near the back wall and rather hard to see. There appear to be no footlockers in this barracks, but the large wall lockers provide ample storage space for clothing and personal possessions. The two lockers left open, at the far end of the room, and the two with doors ajar at right foreground, give the viewer some idea of the kinds of things soldiers needed storage space for. Overcoats and a good bit of equipment hang outside the lockers. At right foreground, we see that the company drummer has stowed his drum and other belongings below the barracks window.

[Negative Number RG-92-F-34-6, National Archives, Washington, D.C.]

Figure 3.43. Cavalry Barracks, Fort Brown, Texas, 1891-1895

Figure 3.43 gives the viewer a first glimpse of the individual mosquito nets issued to each enlisted man as standard equipment. These nets were widely used at all arm posts, particularly those near large bodies of water and in southern climates. Photographs of the nets actually in place and in use are very rare. This one shows the individual nets above each bunk secured on an overhead wire, instead of being supported by a T-bar. The nets are mentioned in detail in the 1884 Quartermaster's Report.

[Negative Number RG-92-F-12-9, National Archives, Washington, D.C.]

Figure 3.44. Artillery Barracks, 4th U.S. Artillery, Battery C, Fort McHenry, Baltimore, Maryland, 1893-1896

The barracks room seen in Figure 3.44 seems to have been readied for a Saturday morning inspection. All footlockers are open, their contents all neatly in order. Bunks are uniformly stripped and all bedding rolled. Dust cloths of printed fabric are all in place at wall storage units. Everything is clean orderly, and in place. The clock on the rear wall registers 10:30 and the bright light coming in the windows tells us it is 10:30 A.M., which probably means that the men are currently outside on parade and being inspected. Once that is completed, the inspecting officer will come inside and check the barracks.

Except for the one Composite No. 10 bunk, the second in the row at the left, all bunks in this photograph are different from those shown earlier. These are a new model introduced between 1893 and 1895, with the plain metal footboard seen here. Also on view are the army's barracks chairs, Model 1884. The stove is the army's cast-iron coal heater No. 6 (or No. 7), Model 1875. The attractive dust covers are made of some sort of civilian fabric, probably an inexpensive cotton print. The overall effect on the room is one of cleanliness, neatness, and uniformity—qualities often overshadowed in the 1800s by strong outcroppings of individuality among the army's healthily unrepressed enlisted men.

[Negative Number RG-92-F-38-5, National Archives, Washington, D.C.]

Figure 3.43

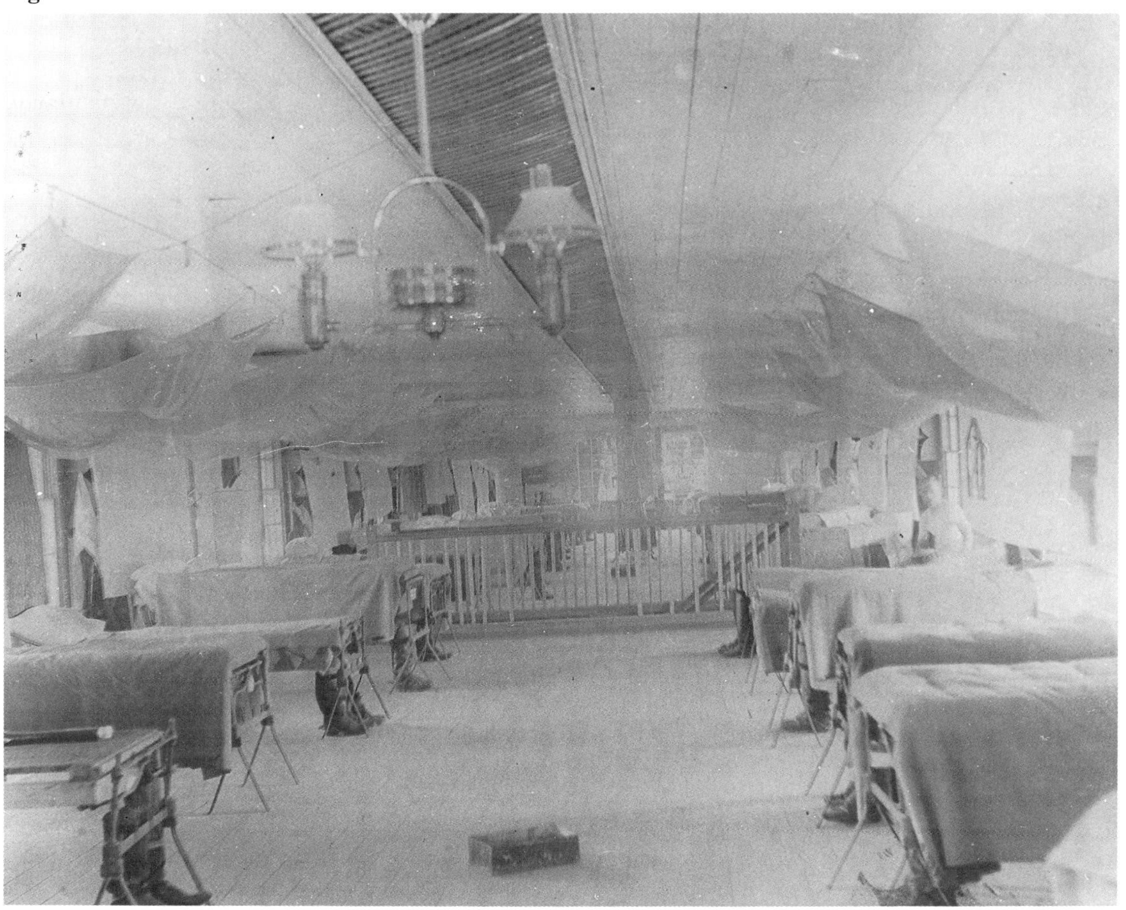

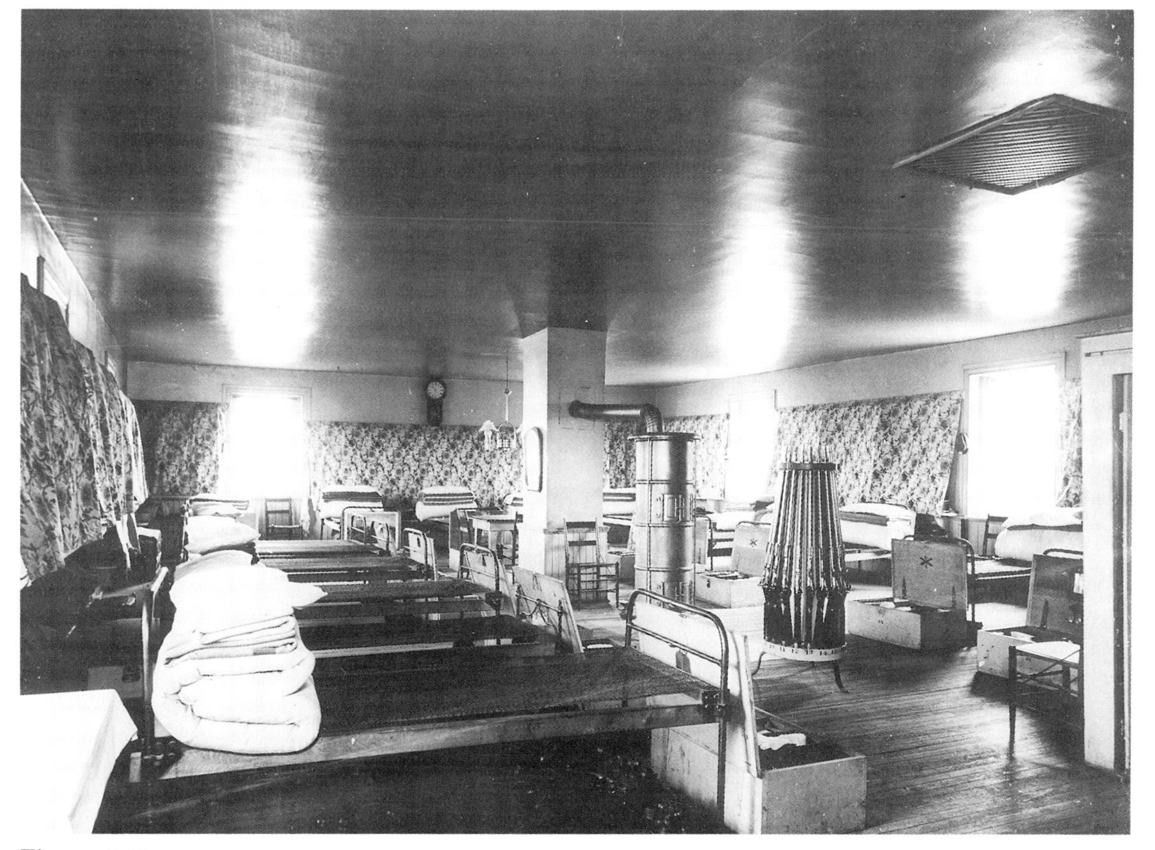

Figure 3.44

Figure 3.45

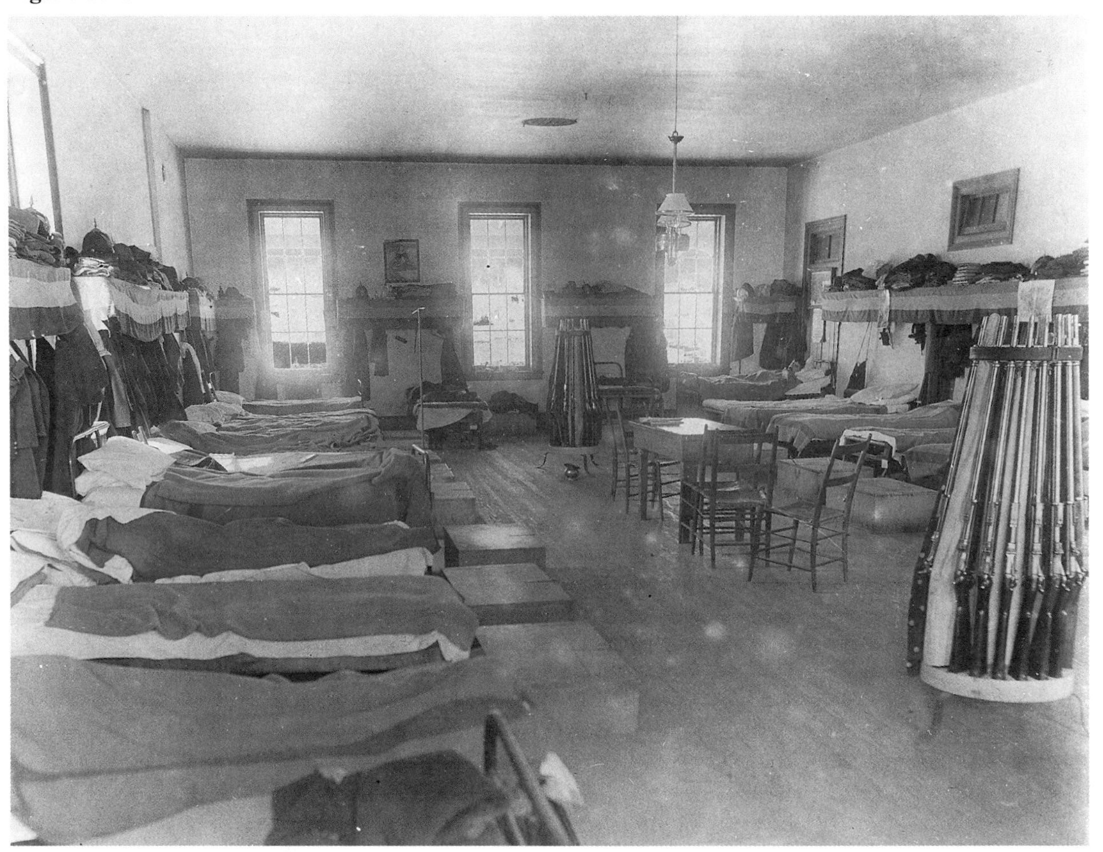

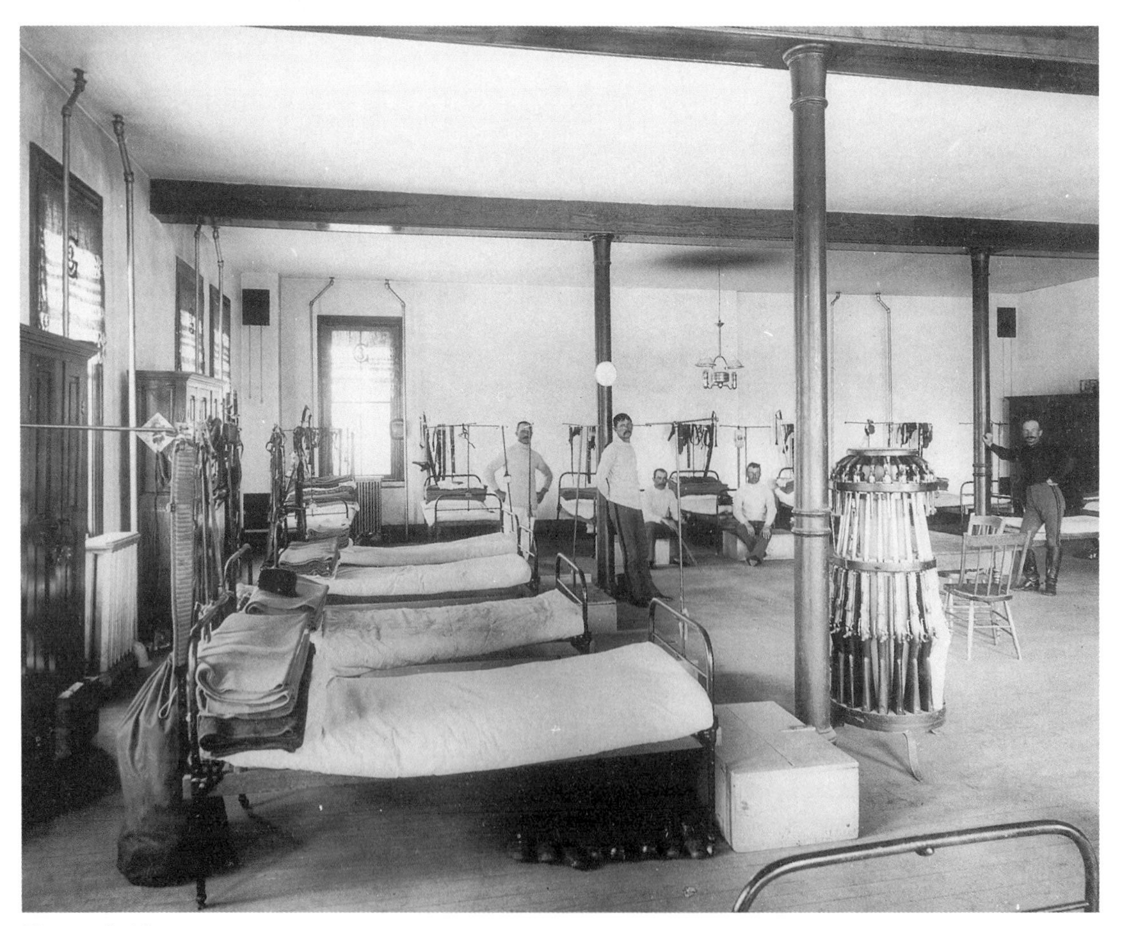

Figure 3.46

Figure 3.45. Infantry Barracks, Fort Hamilton, New York, 1893-1896

In vivid contrast to Figure 3.44, the barracks room in Figure 3.45 is decidedly not ready for inspection. Bunks are rumpled or carelessly spread up, and at least two of these bunks—the one at the window (far right) and the fourth from the front at left—are still occupied. Clothing is stacked haphazardly on wall shelves, coats hang any old way beneath the shelves. A half-closed footlocker at the far end of the room looks as if someone, surprised in the act of rifling the contents, had fled in careless haste. And towels or small articles of clothing hang on a makeshift clothesline in the window at left.

A few less derogatory things can be found in Figure 3.45. The red-white-and-blue bunting that decorates the wall shelves (our imagination tells us that those are the colors in this bit of decoration) is a nice touch. Most of the bunks are the new, cleanly designed model adopted between 1893 and 1895. And some of the rifles on the two gun racks have neat, individual dust covers.

[Negative Number RG-92-F-25-4, National Archives, Washington, D.C.]

Figure 3.46. Artillery Barracks, 3rd U.S. Artillery, Company G, Fort Ethan Allen, Vermont, 1893-1896

As army barracks go, the artillerymen's barracks room in Figure 3.46 seems unusually spacious and uncluttered, and full-sized wall lockers at the heads of the beds along the room's side walls provided plenty of room for the men's clothing and personal belongings. The company insignia—the letter "G" encircling a small "3"—can be seen on the striped window coverings at near left and at the back of the room if one looks very closely. The boots and shoes lined up under the bed in the foreground indicate that placement was then becoming standard practice for the soldiers' footgear. It was still being done that way when I was in the service in the 1950s. The odd-looking small black rectangular shapes, high in both corners of the rear wall, are ventilators. They could be opened or closed by pulling one or the other of the two cords dangling from each ventilator box.

[Negative Number RG-92-F-23-6, National Archives, Washington, D.C.]

Figure 3.47, 3.48, and 3.49. Barracks Buildings, 1893-1896: Fort Bliss and Fort Sam Houston, Texas (Infantry Barracks); Jackson Barracks, Louisiana (Cavalry Barracks—Musicians)

All three barracks in Figures 3.47, 3.48, and 3.49 show the army's new model bunk with the neat, uncluttered metal footboard. And lengths of tent canvas—"shelter halves"—appear as bed covers on the bunks in Figures 3.48 and 3.49. Figure 3.48 also has an excellent view of the army's older barracks lamp, the tall Model 1881.

Regulation footlockers appear in Figures 3.47 and 3.49, and in Figure 3.48 we see what appear to be closed wall lockers—five each at both left and right against the rear wall. All these rooms probably have good-sized wall lockers at some point that were out of camera range when the photograph was made.

[Negative Numbers RG-92-F-9-5; RG-92-F-56-11; and RG-92-F-30-9, National Archives, Washington, D.C.]

Figure 3.47

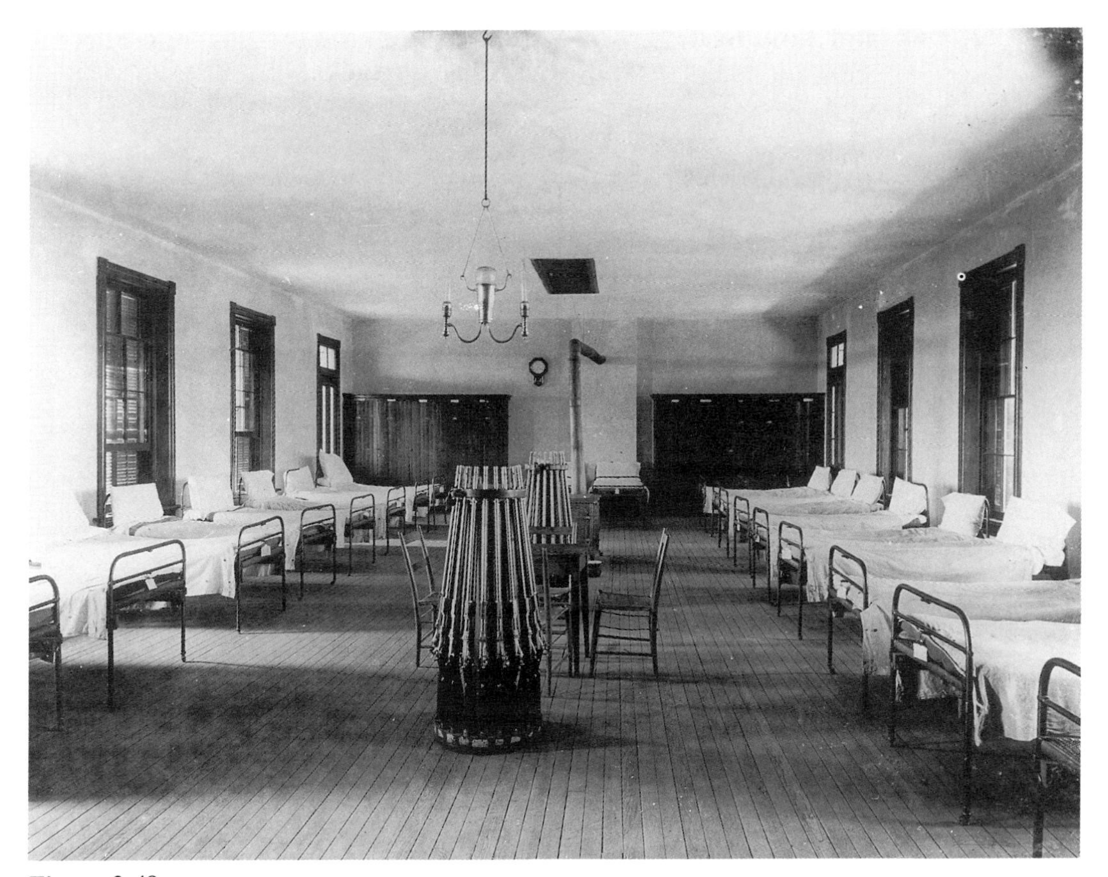

Figure 3.48

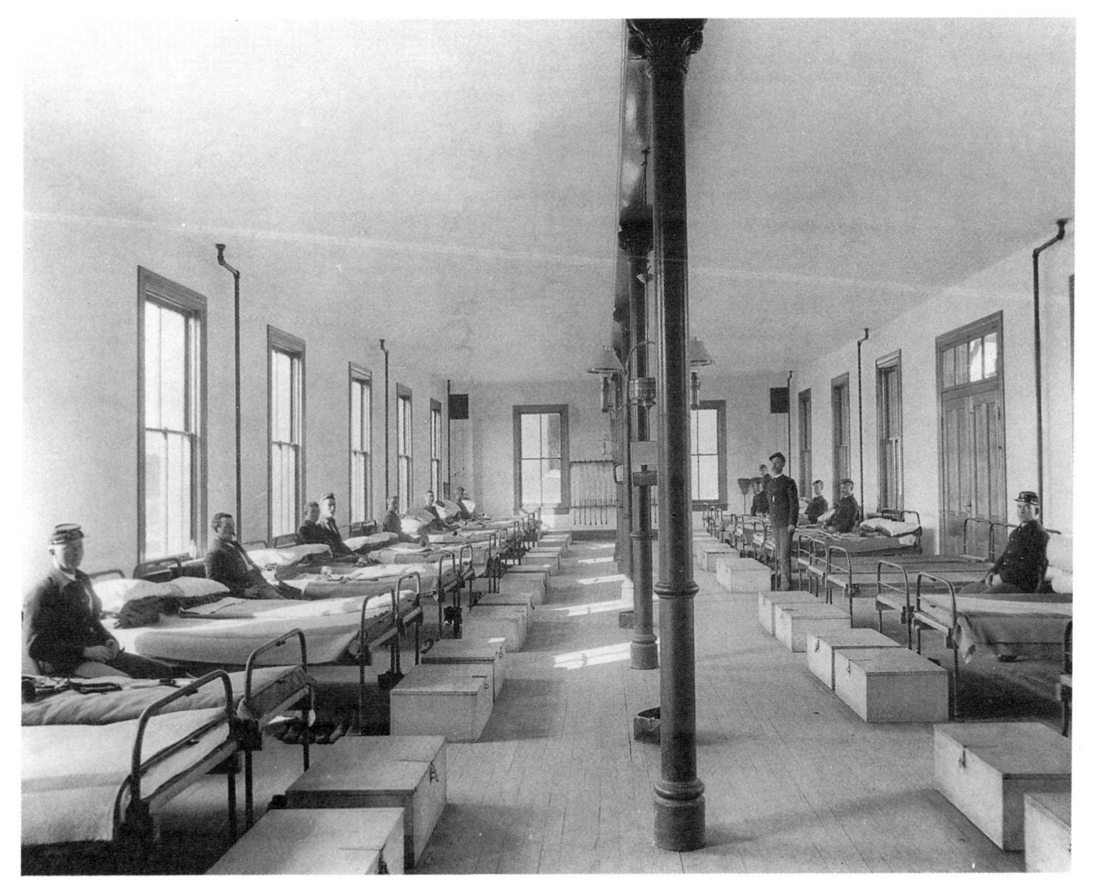

Figure 3.49

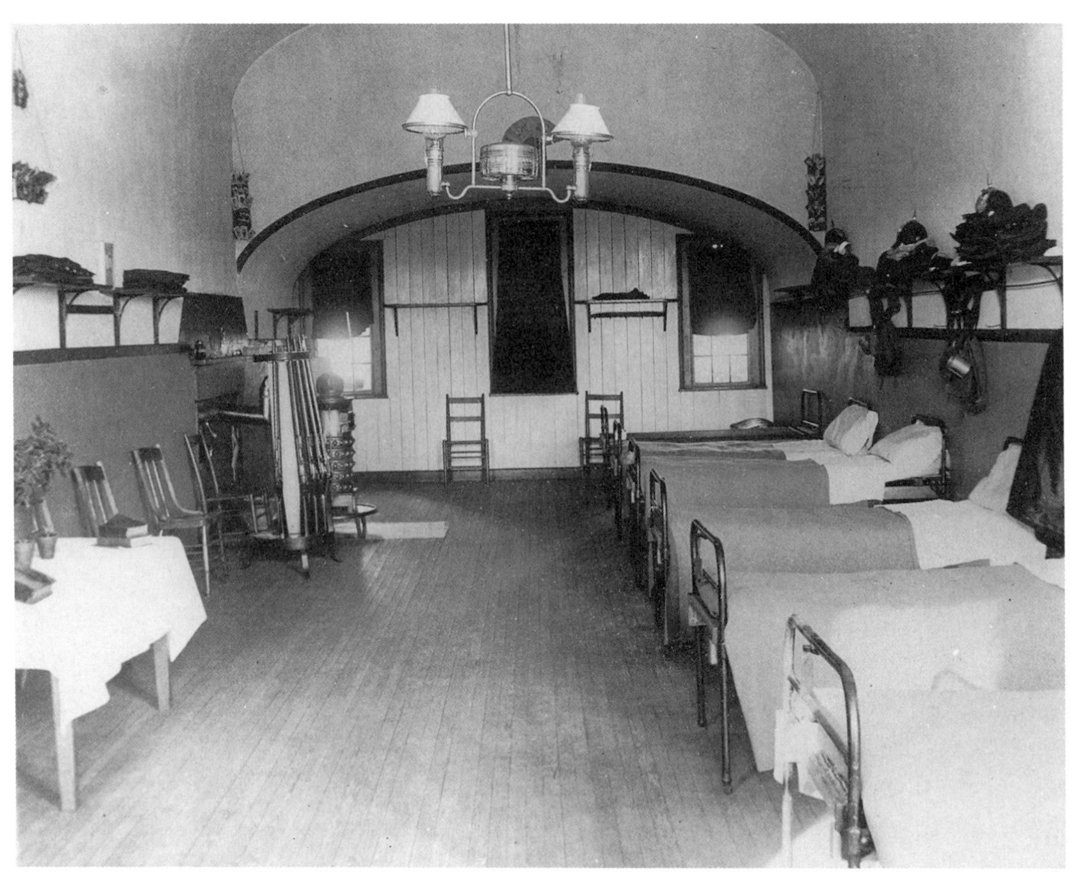

Figure 3.50

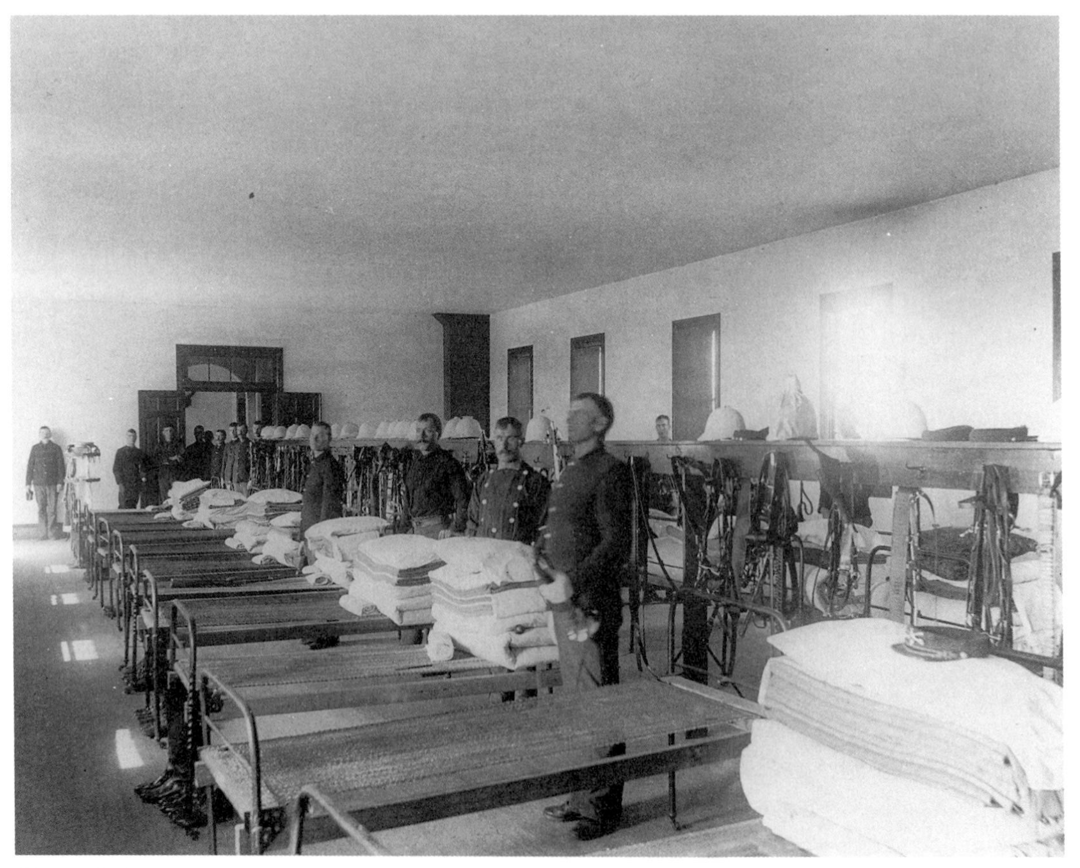

Figure 3.51

Figure 3.50. Artillery Barracks in a Casemate, Fort Warren, Massachusetts, 1893-1896

In Chapter 1 of this book, Figures 1.2 and 1.3 present two views of casemate quarters for U.S. Army officers; Figure 3.50 is the only photograph in this book showing casemate quarters for enlisted men. As noted earlier, in the caption for figure 1.2, casemates were fortress chambers, often vaulted, and always very well secured. Casemates were usually built to serve as artillery batteries, although they could serve as barracks buildings, with some adaptations. The U.S. Surgeon General wrote of casemates in his report dated December 5, 1870:

> Casemates, if constructed with a view to their greatest efficiency as casemates, are unfit quarters, by reason of dampness, darkness, and insufficient ventilation if kept improperly warm; and it would, no doubt, be good economy to discontinue their use as quarters and lodge the men in proper barracks, even if these had to be destroyed, in the case of actual hostilities, every ten or fifteen years to meet the exigencies of the service.

However, that may be, there remained a few forts in which the army continued housing its men in casemates, and Fort Warren was one of them.

The stove in this barracks room, seen between the gun rack and the window at left rear, is patently non-regulation with its odd five rows of what look like knobs circling the lower part of it. The room decor has other individualistic touches: a fan spread open above the arch, at both right and left, and the arched top of a civilian trunk seen at the rear wall, just beyond the last bunk on the right.

No footlockers and no wall lockers are seen in the photograph, but very likely there were wall lockers in the area behind the photographer.

[Negative Number RG-92-F-74-4, National Archives, Washington, D.C.]

Figure 3.51. Cavalry Barracks, Fort Meyers, Virginia, 1893-1896

The viewer will notice immediately the two most unusual things about the cavalry barracks in Figure 3.51: the bunks are aligned down the center of the room, instead of backed up with the headboards against a wall, and instead of the usual storage shelves or lockers on the wall behind the bunks, that space is given over to an unusual free-standing rack for storing the men's headgear and the horses bridles and harnesses. Wall lockers may very well have been provided in an area of the room not photographed.

The viewer forms an impression that this room is about to be inspected—perhaps because of its neat appearance and because most of the men seem just at the point of standing at attention, as is the company's bugler in the near right foreground. Further, the total absence of weapons from the circular gun rack at the far end of the room could indicate that an inspection or a parade is currently in progress outside, with barracks inspection soon to follow.

[Negative Number RG-92-F-45-7, National Archives, Washington, D.C.]

Figure 3.52. Infantry Barracks, Plattsburg, New York, 1893-1896

The Plattsburg, New York infantry barracks, photographed here in the last decade of the nineteenth century, was a very up-to-date barracks room, fitted with wall lockers, the latest models of the army's lamp and bunks—all regulation equipment, as is the shining spittoon of treated fiberboard in the immediate foreground.

[Negative Number RG-92-F-47-6, National Archives, Washington, D.C.]

Figure 3.53. Unidentified Army Post, 1905-1910

Figure 3.53 illustrates a view of an unidentified turn of the century army post which is included to show the new Quartermaster bunk, adopted as army equipment in 1905. It was, at that time, generally used only in hospitals and cadet barracks, and although it was introduced in 1905, it may have been manufactured as early as 1897. The viewer who has looked closely at the army bunks shown thus far will notice that the woven-wire bedsprings of the older bunks have been superseded here by a modern-day set of wire springs attached to the bed frame by a short section of wire coils designed to give a bed more "bounce." The same viewer will probably also wonder why the sergeant seated in dress uniform (right foreground) is wearing what looks like argyle socks.

[William Henry Collection, St. Louis, Missouri, 1905-1910]

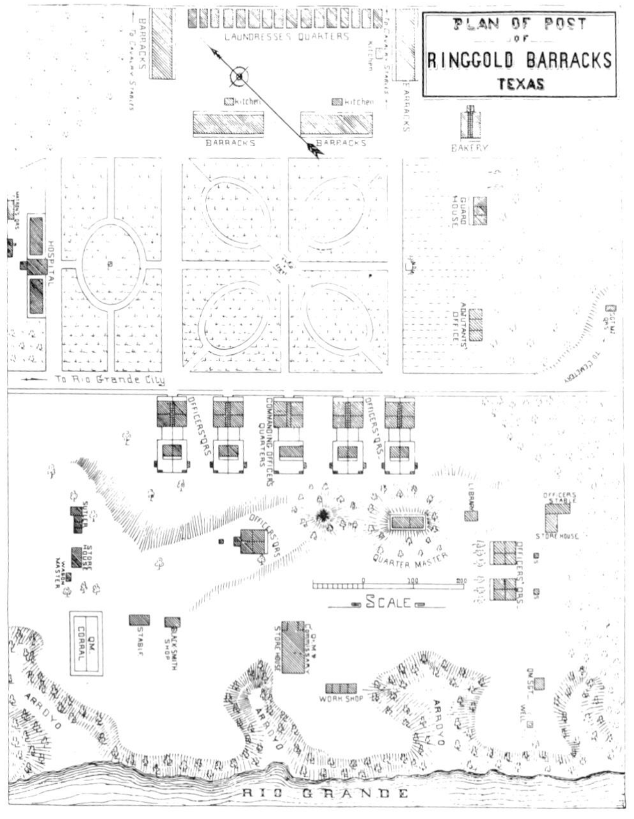

From Circular No. 8, *Report on Hygiene of the U.S. Army.*

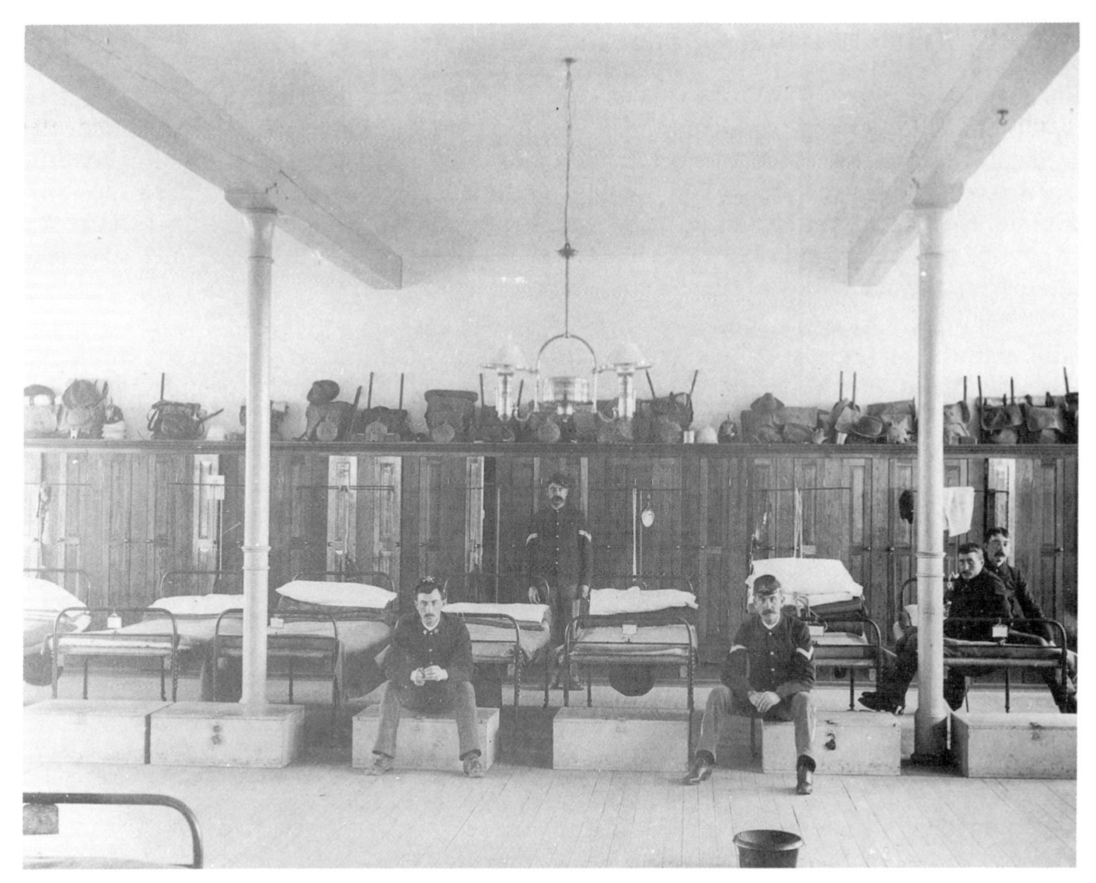

Figure 3.52

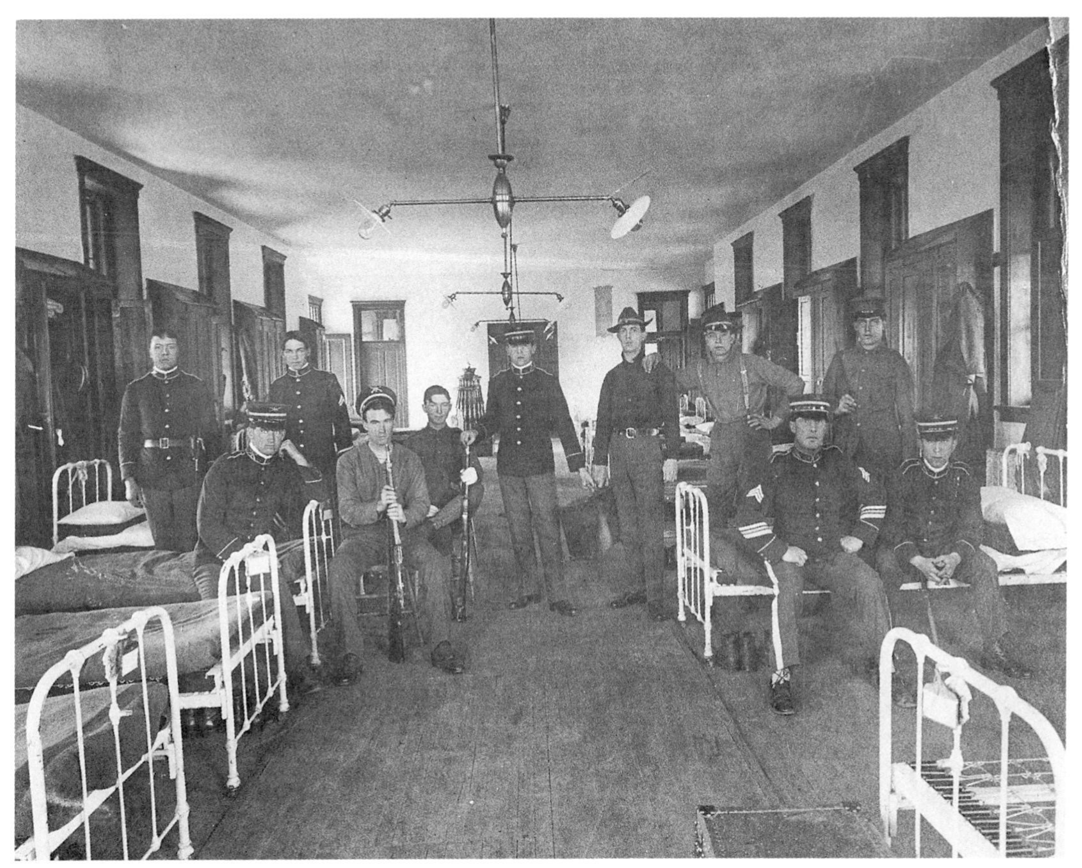

Figure 3.53

Figure 3.54. 2nd Cavalry Bandsmen, Fort Wingate, New Mexico, 1894-1896

Each company or troop in the nineteenth century had its regular field musicians—a drummer, a fifer, or a bugler. Apart from the field musicians were the company bandsmen. Only a regimental headquarters maintained a military band. The bandsmen, like the company drummer, fifer, and bugler, generally had specialized duties to perform that required specialized discipline, equipment, and training; therefore, these men were not usually held to the standard routine duties required of most men in the ranks. Bandsmen, usually assigned to quarters separate from the main body of regular soldiers, often enjoyed more privileges than the average enlisted man.

Although the regimental colonel commanding a post had a great deal to say about how much individuality might be allowed in personal dress and barracks decoration throughout the regiment, the post's bandsmen often enjoyed much leeway in both areas. Here, in Figure 3.54, the wall decorations in the bandsmen's quarters are highly individualistic. Indian artifacts, textile and photographic prints, animal horns, and musical instruments crowd the wall in front of which these four men pose. The parlor-like table setting in front of the window at far right and the window drapery set a different tone from that of the average military barracks. And the two men at left, in their timeless turtlenecks, might have just come in from moonlighting as photographers' models for a 1890s clothing catalog.

[Negative Number RG-495, Louis Dunston Collection, National Archives, Washington, D.C.]

Figure 3.54

Figure 3.55. Infantry Barracks, Fort Shaw, Montana, 1890-1892

The individualism of the average nineteenth-century infantryman is well-illustrated in the decor of the infantry barracks seen in Figure 3.61. Prints and photographs appear on the walls, and decorative matched valances have been used to dress the windows. This room also was provided with somewhat unusual wall racks that hold the men's rifles in a vertical position. One soldier's rifle can be seen, so mounted, on a wall (left foreground). The three rifles in the tripod arrangement in the center of the room have been placed there merely as a decorative touch, either by or for the photographer.

[Negative Number 947-376, Montana Historical Society, Helena, Montana. Photograph by C.E. Lemunyon.]

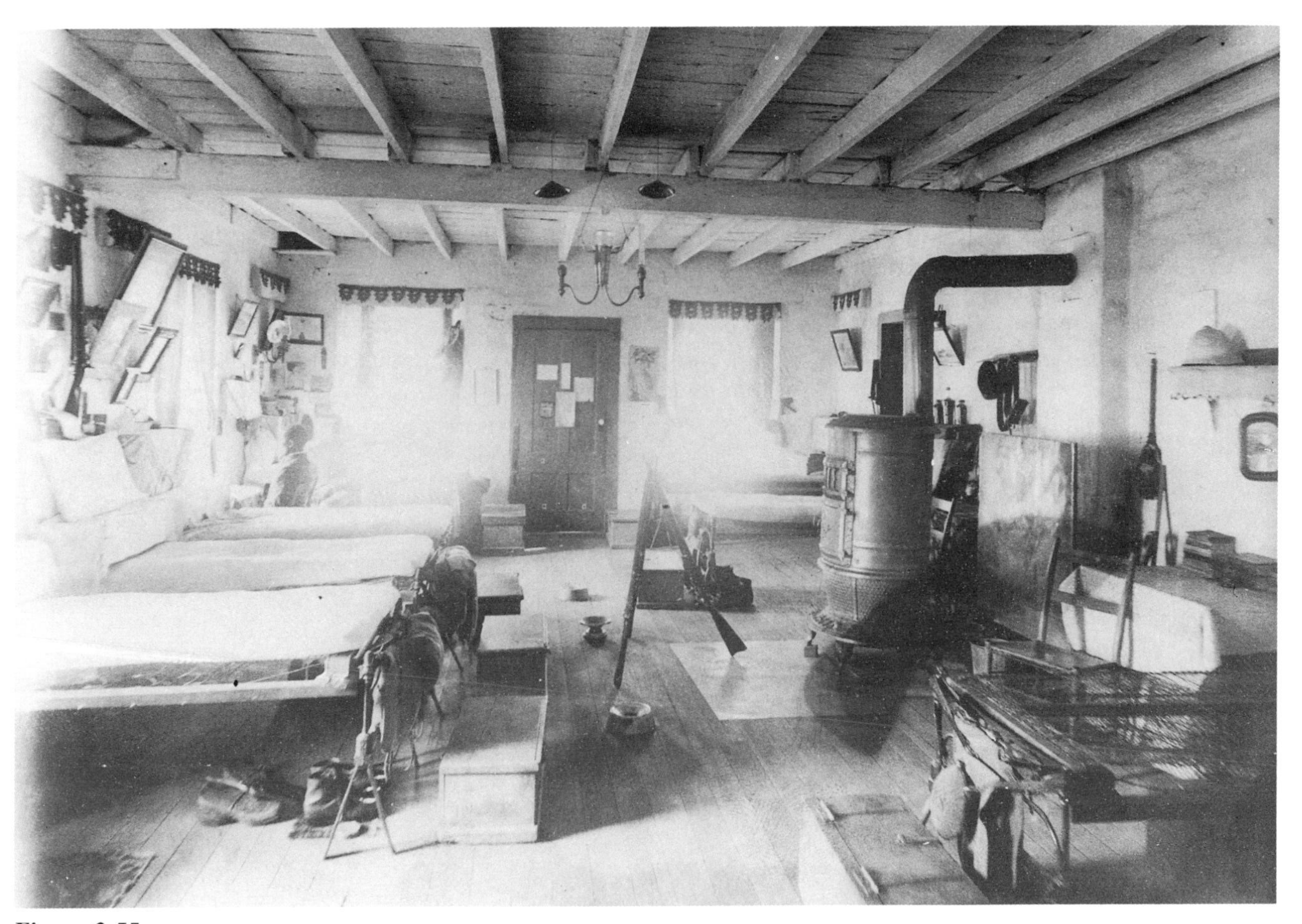

Figure 3.55

Figure 3.56. Infantry Barracks, Company K, 23rd Regiment, Fort Mackinac, Michigan, ca. 1886

This is the first floor squad room of the soldiers' barracks, still standing at the Fort. The table on which two soldiers are playing cards is marked on the apron QMD, thus finding it as Quartermaster Department issue furniture. QMD is quite rare, and this is an excellent view of the table. The chairs also appear to be regulation type. The Composite iron bunks and one Snead Bunk, at the rear left have all been fitted with the wire woven mattress frames. Post surgeon, Dr. John R. Bailey remarked in May 1885, "The enlisted men have been supplied with one mattress each, stuffed with cotton...one cotton pillow each, two cotton pillow cases, and four sheets to each bed. Such accommodations are very much better than those furnished by ordinary hotels, and may be said to be truly luxurious."

Also of interest is the bulletin board on the rear wall, and what appear to be training/fencing bayonets on three of the rifles in the gun rack. Spit boxes have been replaced by the new metal spittoons. Truly luxurious.

[Mackinac State Historic Park, Violet Bowling Collection]

Figures 3.57 and 3.58. Two Cavalry Barracks, Fort Robinson, Nebraska, 1891-1895

Compare these two enlisted men's barracks at the same post; one of the barracks houses a black company. The second interior may house another company from the same black regiment. Except for different model bunks, stoves, and lamps, the arrangement is nearly identical, including the presence of a pool table at the end of each room. Non-commissioned officers' quarters are probably located behind the board partition at the end of one barrack.

[Negative Number RG-92-F-54-7, National Archives, Washington, D.C.]

Figure 3.56

Figure 3.57

Figure 3.58

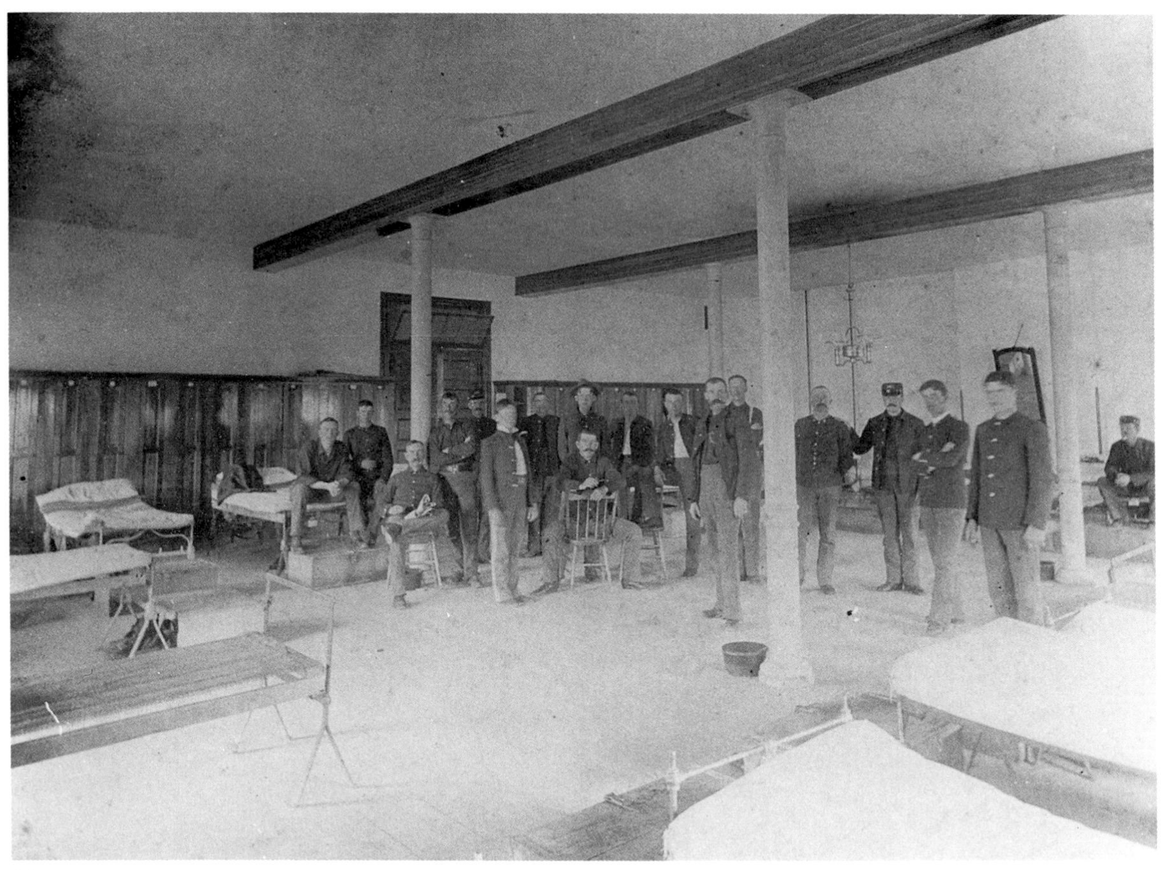

Figure 3.59

Figure 3.60

Figure 3.59. Infantry Barracks, Fort Logan, Colorado, Marched Company G, Lower Squad Room, 1896-1902

Of special interest is the large mirror on the wall implying a greater attention to how soldiers appeared on Parade. Also the mixture of bunks, even a few of the Snead Models. Wall lockers now hide all the uniforms and equipment. Note also how the wire-woven bunk bottoms are attached differently to the same model bunks.

[Jerome Greene Collection]

Figure 3.60. Cavalry Barracks, Unknown location, probably in the southwest, 1892-1895

This one story, high ceiling room is very similar to the surviving barracks at Fort Davis, Texas. All of the bunks appear to be the Model No. 10 Composite with the wire woven mattress. The stoves, chairs, and foot lockers are regulation, as is the Model 1891 Barrack Lamp. Note the bulletin board on the far wall.

Also of interest is the close-up view of the field gear. The belt has 10 cartridges in it and the tin cup is slung on the canteen strap. Canvas garters lie on the bottom of the bed, and the coat lying on the pillow appears to be a jacket as opposed to a 5-button coat.

[Smithsonian Institution]

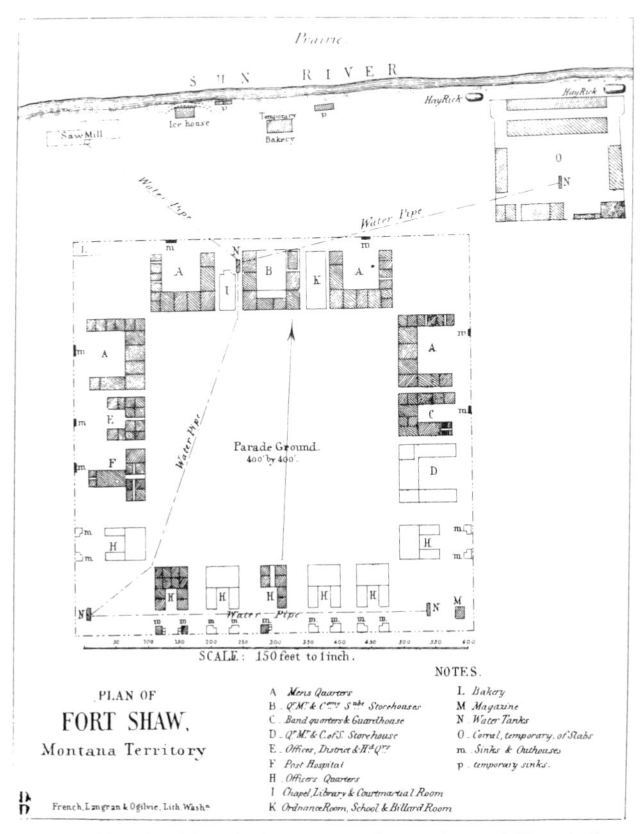

From Circular No. 4, *Report on Barracks and Hospitals.*

A Note on Figures 3.61 through 3.81

The last twenty-one photographs in Chapter 3 have been included to complete this showing of the range of views of nineteenth-century army enlisted men's barracks kept in the collections of the National Archives. Although these last photos may not add a great deal of new data to the specific information evident in the preceding photographs, they do add to the viewer's general knowledge and feeling for the army of the 1800s. To have left out these final pictures would have, I believe, lessened the comprehensive scope of this book. Figures 3.61 through 3.81 appear with identifying titles only. I have offered no comment on any specific item in the photographs that conclude this section. I hope that the viewer will find each of them as interesting and informative as those that precede and follow them.

Figure 3.61. Engineers Barracks, Location Unknown, From a Painting by A. Cast, 1894-1897
[Negative Number NA-111-SC-100081, National Archives, Washington, D.C.]

Figure 3.62. Barracks, Fort Adams, Rhode Island, 1891-1895
[Negative Number RG-92-F-1-8, National Archives, Washington, D.C.]

Figure 3.63. Barracks, Fort Alcatraz, California, 1891-1895
[Negative Number RG-92-F-2-4, National Archives, Washington, D.C.]

Figure 3.61

Figure 3.62

Figure 3.63

Figure 3.64. Barracks, 3rd U.S. Infantry, Company A, Fort Barrancas, Florida, 1891-1895
[Negative Number RG-92-F-6-8, National Archives, Washington, D.C.]

Figure 3.65. Infantry Barracks, Fort Bayard, New Mexico, 1891-1895
[Negative Number RG-92-F-7-7, National Archives, Washington, D.C.]

Figure 3.66. Infantry Barracks, Columbus, Ohio, 1891-1895
[Negative Number RG-92-F-16-11, National Archives, Washington, D.C.]

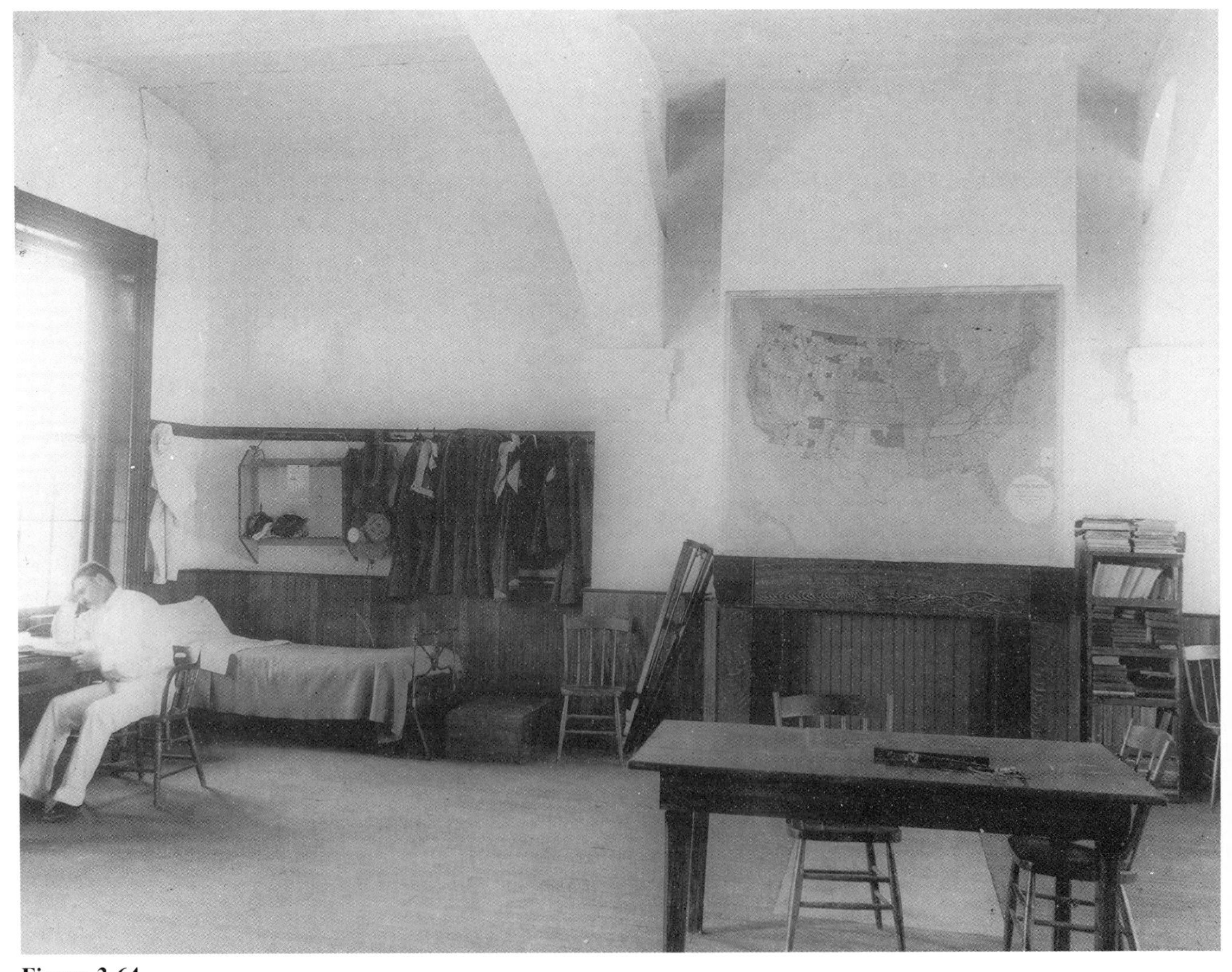

Figure 3.64

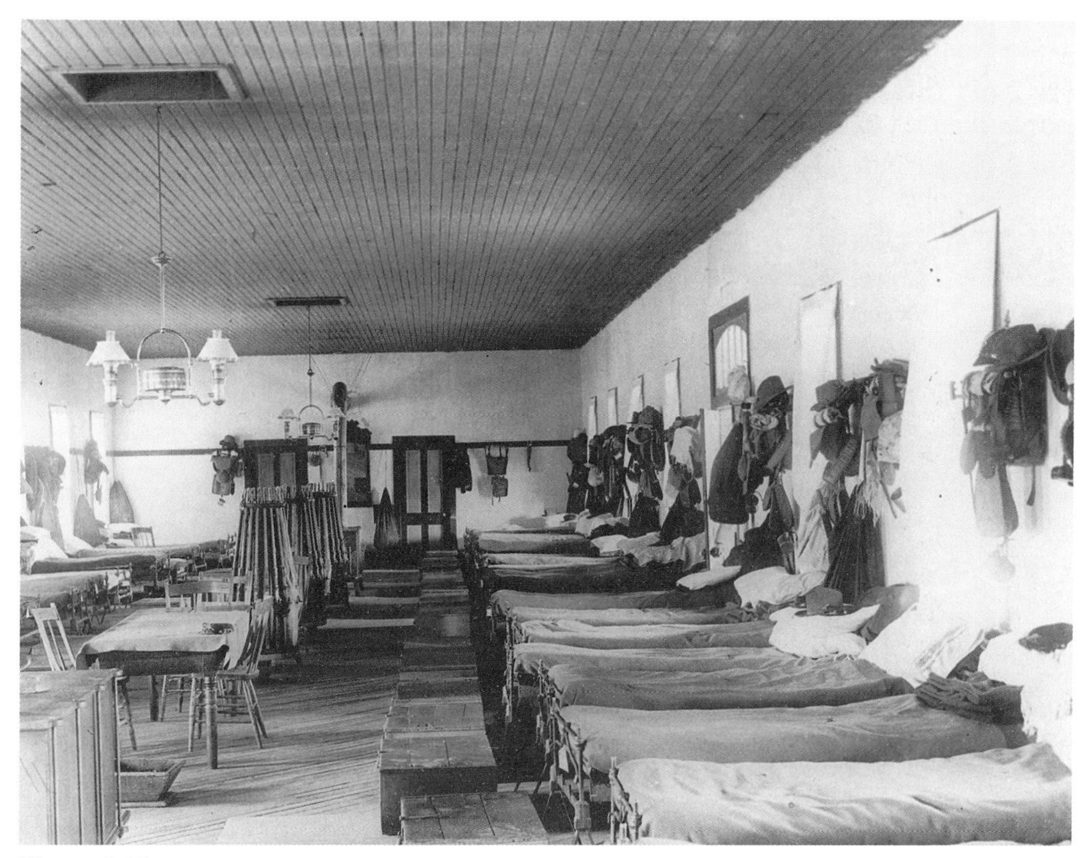

Figure 3.65

Figure 3.66

Figure 3.67. Infantry Barracks, Columbus, Ohio, 1891-1895
[Negative Number RG-92-F-16-14, National Archives, Washington, D.C.]

Figure 3.68. Infantry Barracks, Columbus, Ohio, 1891-1895
[Negative Number RG-92-F-16-12, National Archives, Washington, D.C.]

Figure 3.69. Infantry Barracks, Fort Columbus, New York, 1891-1895
[Negative Number RG-92-F-15-6, National Archives, Washington, D.C.]

Figure 3.67

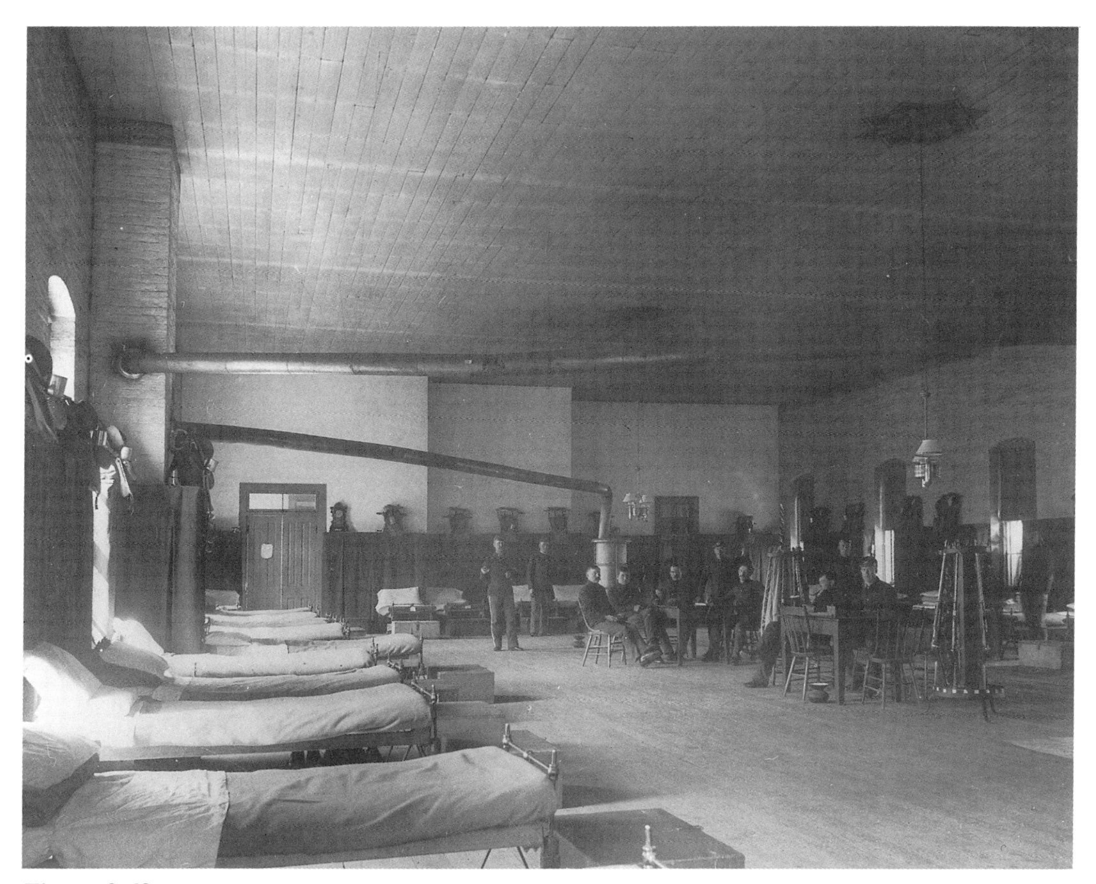

Figure 3.68

Figure 3.69

Figure 3.70. Infantry Barracks, Fort D.A. Russell, Wyoming, 1892-1895
[Negative Number RG-92-F-18-10, National Archives, Washington, D.C.]

Figure 3.71. Cavalry Barracks, Fort D.A. Russell, Wyoming, 1895-1890
[William Henry Collection, St. Louis, Missouri]

Figure 3.72. Artillery Barracks, Fort Monroe, Virginia, 1891-1895
[Negative Number RG-92-F-44A-10, National Archives, Washington, D.C.]

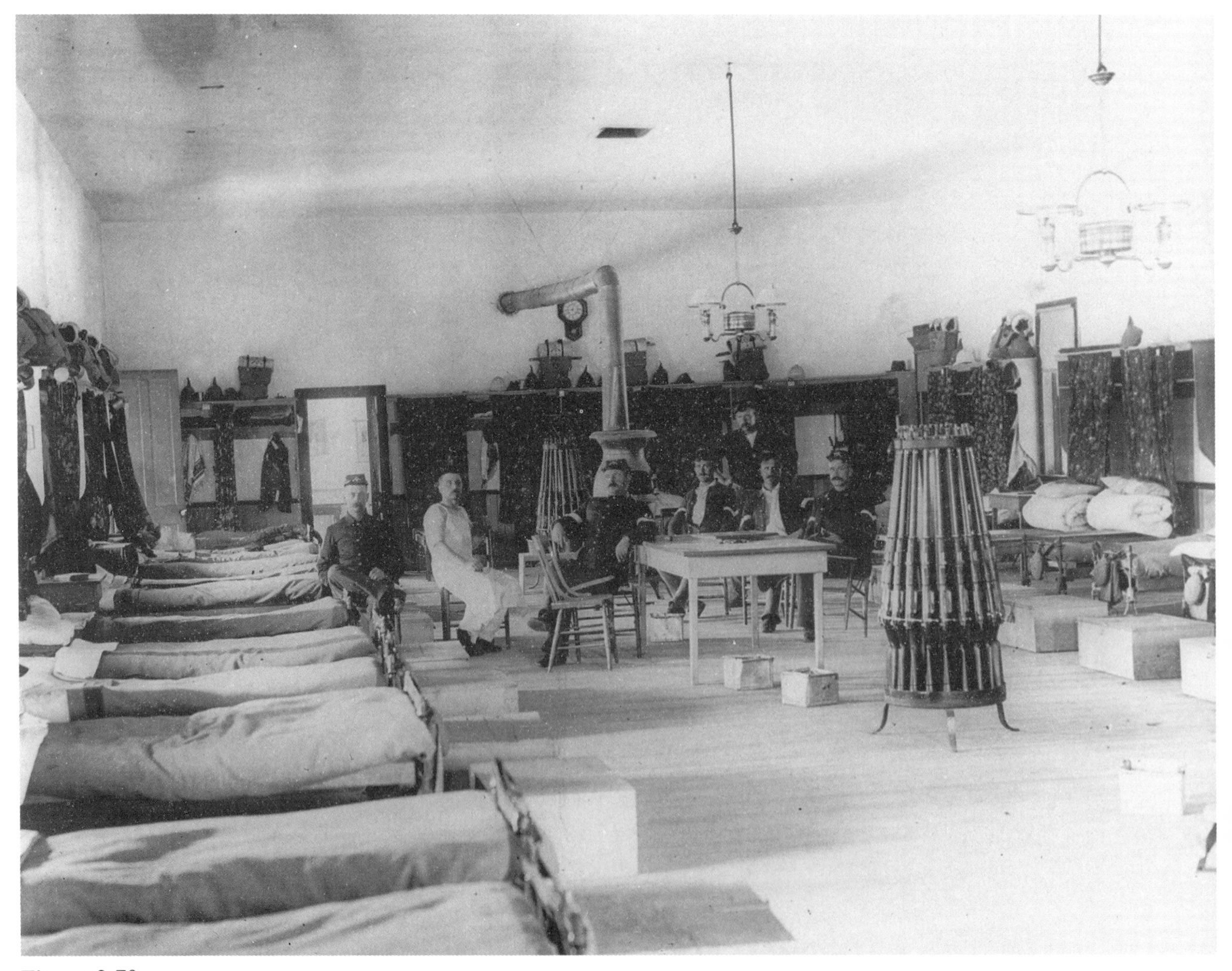

Figure 3.70

Figure 3.71

Figure 3.72

Figure 3.73. Infantry Barracks, Mount Vernon, Alabama, 1891-1895
[Negative Number RG-92-F-44B-27, National Archives, Washington, D.C.]

Figure 3.74. Infantry Barracks, Fort Porter, New York, 1891-1895
[Negative Number RG-92-F-48-7, National Archives, Washington, D.C.]

Figure 3.75. Barracks, Fort Preble, Maine, 1891-1895
[Negative Number RG-92-F-49-3, National Archives, Washington, D.C.]

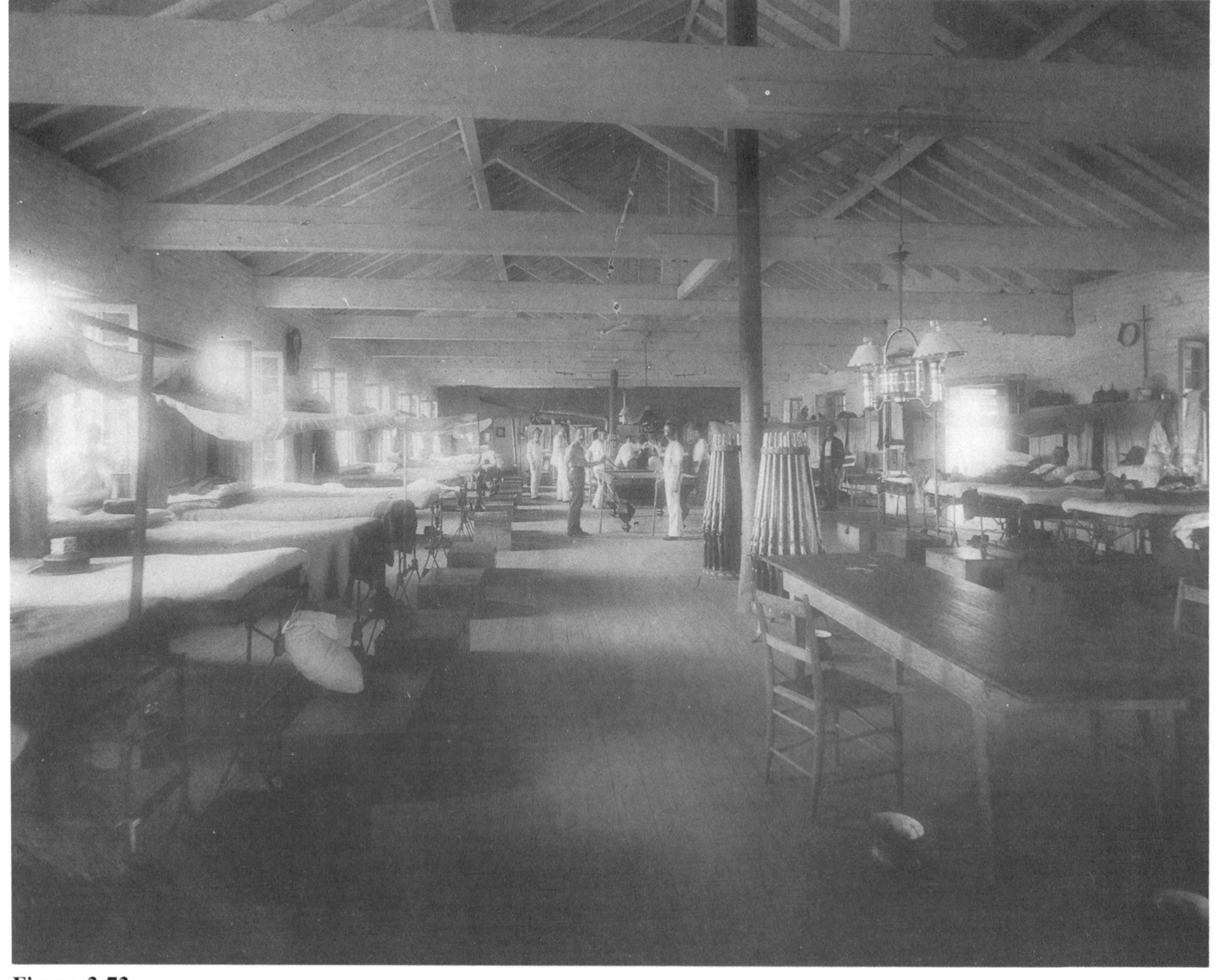

Figure 3.73

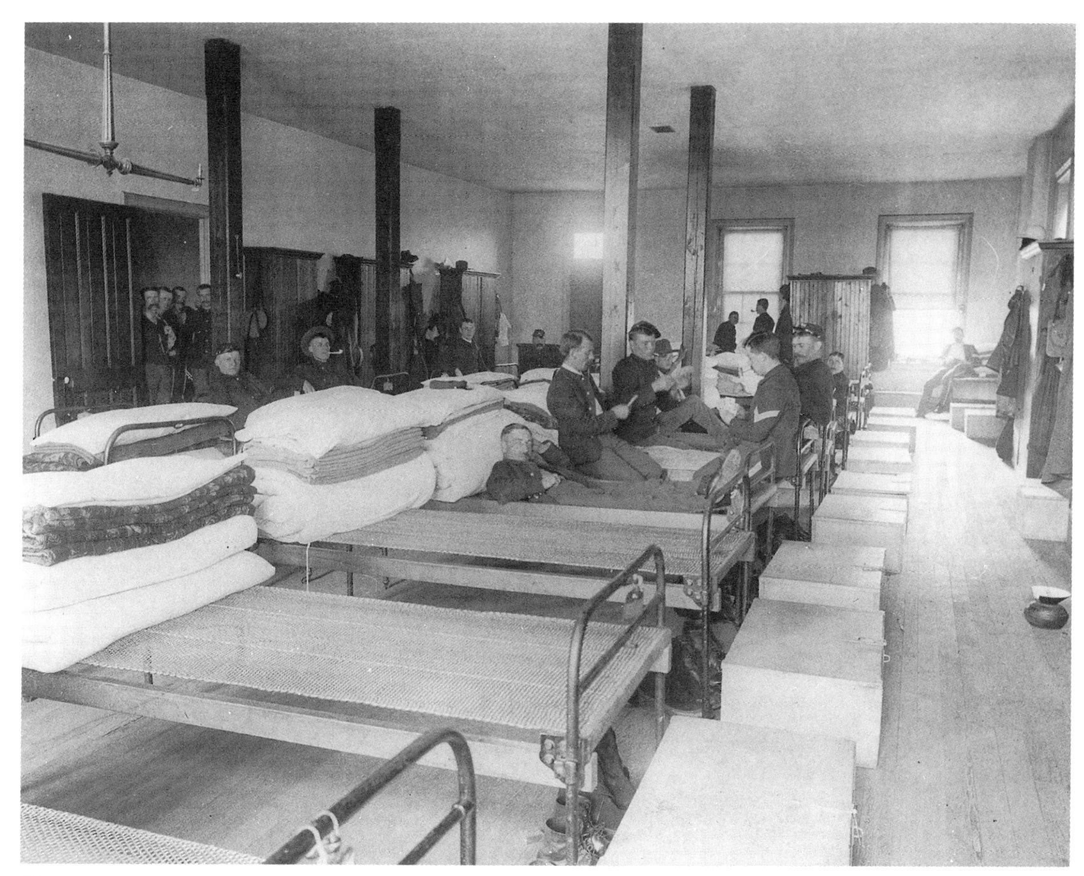

Figure 3.74

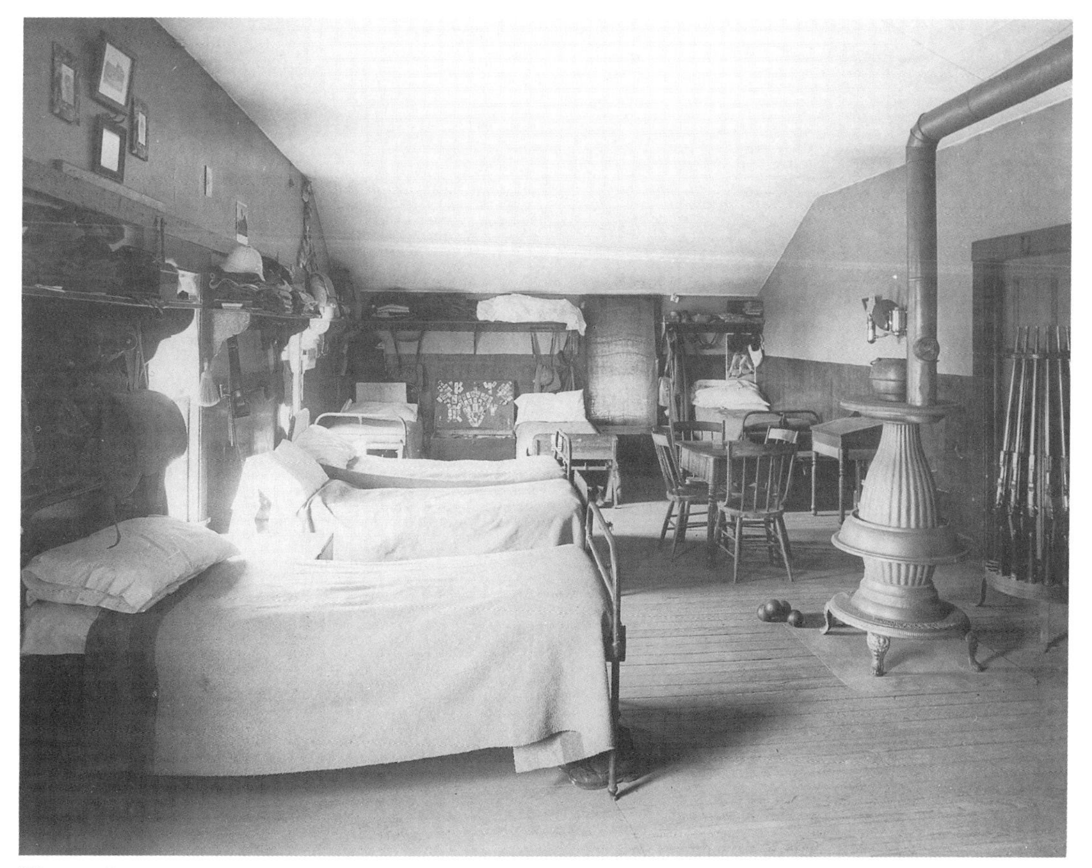

Figure 3.75

Figure 3.76. Infantry Barracks, Fort Snelling, Minnesota, 1891-1895
[Negative Number RG-92-F-65-6, National Archives, Washington, D.C

Figure 3.77. Washington Barracks, Washington, D.C., 1891-1895
[Negative Number RG-92-F-76-7, National Archives, Washington, D.C.]

Figure 3.78. Infantry Barracks, Fort Wayne, Michigan, 1891-1895
[Negative Number RG-92-F-77-7, National Archives, Washington, D.C.]

Figure 3.76

Figure 3.77

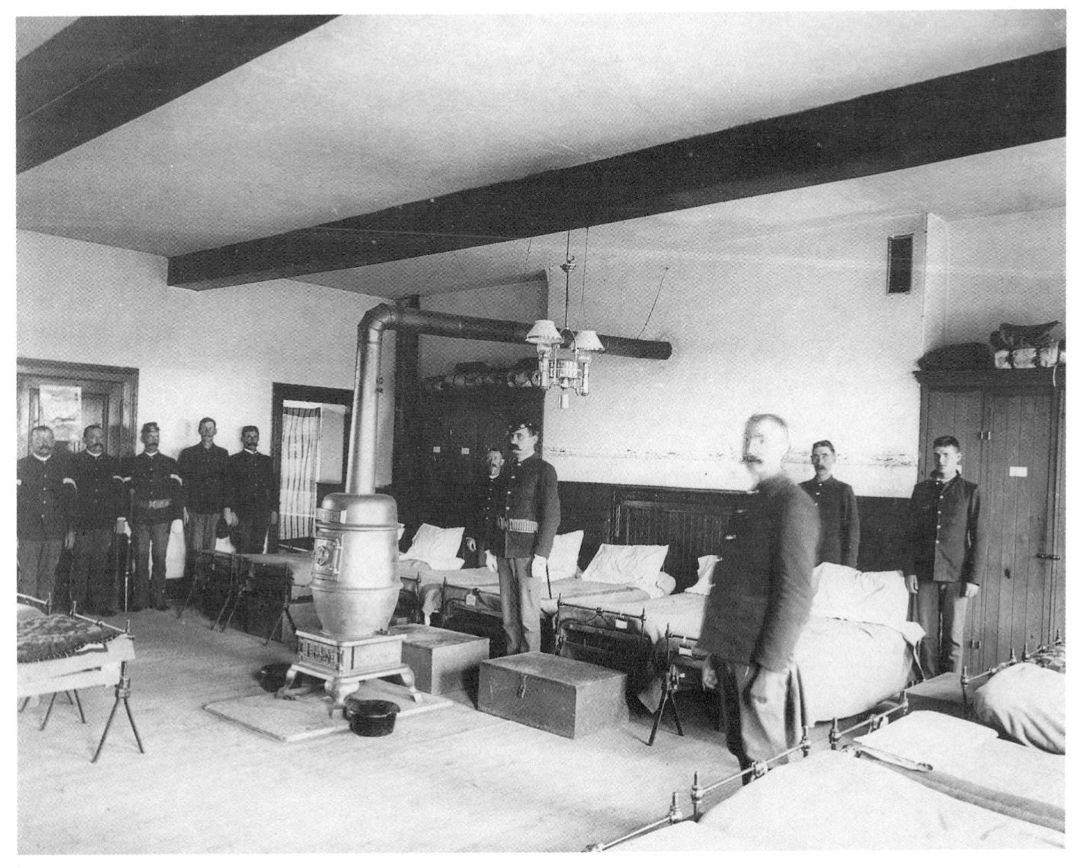
Figure 3.78

Figure 3.79. Barracks, Willets Point, New York, 1891-1895
[Negative Number RG-92-F-80-10, National Archives, Washington, D.C.]

Figure 3.80. Cavalry Barracks, Fort Wingate, New York, 1891-1895
[Negative Number RG-92-F-81-8, National Archives, Washington, D.C.]

Figure 3.81. Infantry Barracks, Fort Wood, New York, 1891-1895
[Negative Number RG-92-F-82-5, National Archives, Washington, D.C]

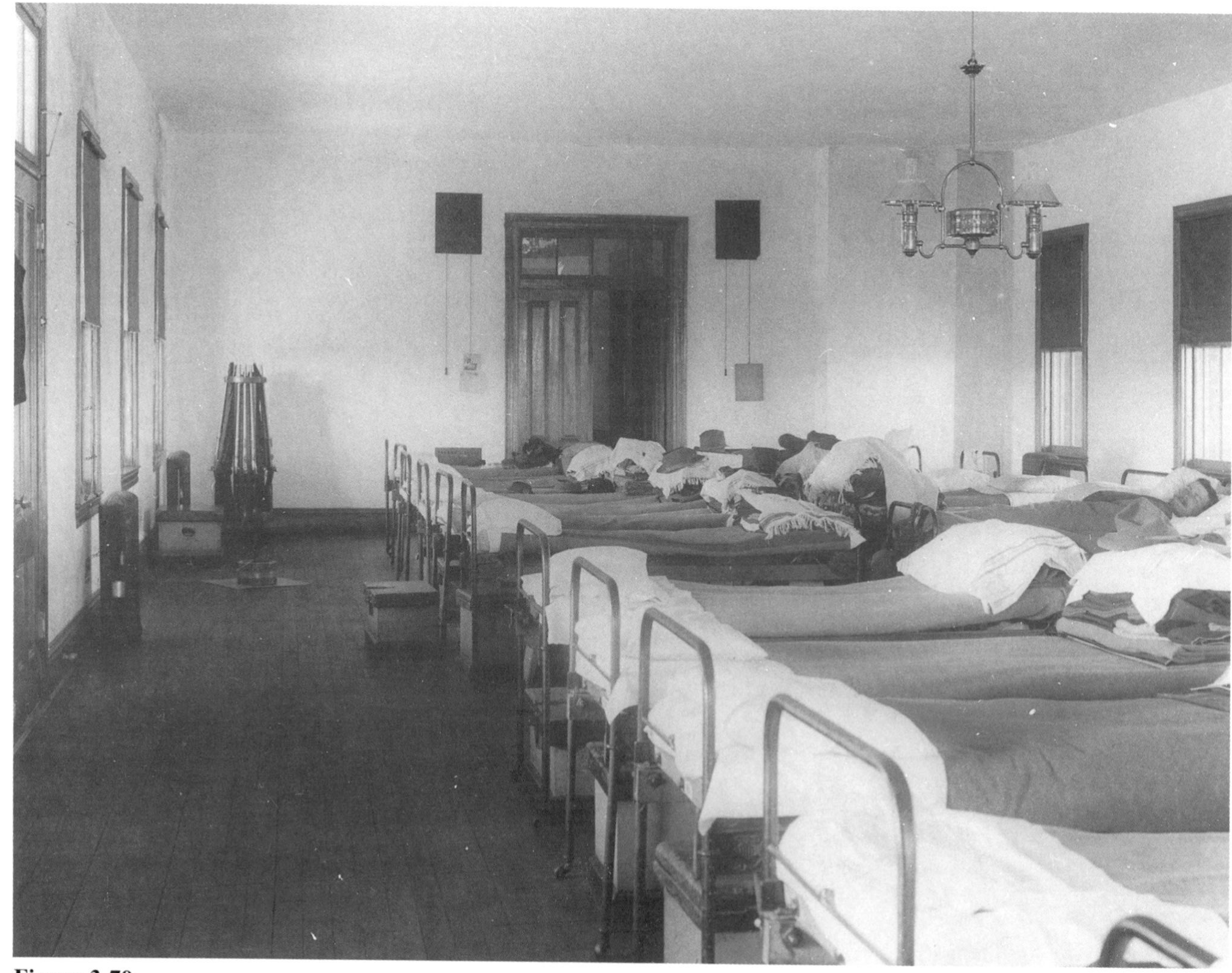

Figure 3.79

Figure 3.80

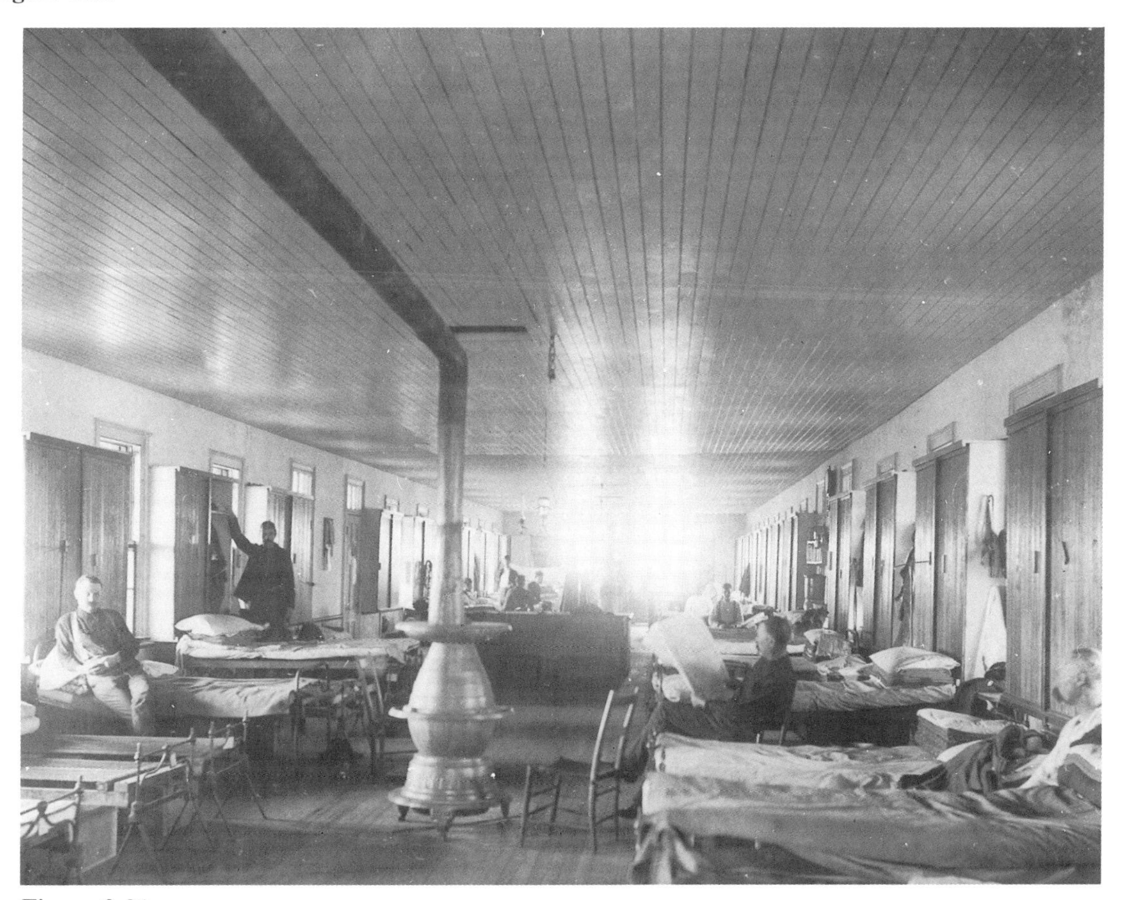

Figure 3.81

MESS HALLS AND KITCHENS

The first major statement that the army circulated for its cooks was a perspective commentary entitled "The Cook's Creed" published originally in Captain James Sanderson's book *Culinary Hints for the Soldier* (1862). One notices that the book's title addresses the *soldier*—not the *cook*—and there was a reason for that: the army did not formally define and establish the position of cook as a permanent military occupation until the late 1870s. During the first two-thirds of the nineteenth century, the job of cooking for a company of soldiers rotated among all enlisted men in the company.

Of that aspect of life for the regular army's average enlisted man on frontier duty in the mid-1800s, authors of *The Story of Great American West* make the following observation:

> The soldiers took turns cooking, so most meals were boiled and bad: stew, hash, baked beans, hardtack (often moldy), salt pork (sometimes maggoty), dried apples and potatoes, and coffee. At forts lacking gardens, the men came down with scurvy. Despite the efforts of post surgeons, sanitation was terrible, and disease took a far higher toll than battles with the Indians.[34]

In the field or on long marches or patrols, a soldier carried in his haversack a day's supply of hardtack and salt pork, the nineteenth-century army's "convenience foods," universally groused about; they were dependable, all but imperishable, and portable. A day's rations—food for two meals—consisted of twelve hardtack "crackers," twelve ounces of salt pork, and a small bag of coffee beans. In addition to carrying one day's subsistence rations, most haversacks also held a small tin plate, knife, fork, and spoon, a few matches in a cartridge case, a twist of tobacco, a straight razor, a sewing kit, a spare pair of socks, and a box of cartridges.[35]

Until past mid-century, at most garrisons, certain of the company members' names appeared and reappeared repeatedly on the duty roster as cooks, until it seemed that those men were being assigned permanent duty as company cook, whether such a position existed or not. As to food preparation, the military diet varied, for better or worse, from post to post. Garrisons with skillful, enthusiastic gardeners had much better food—and health—than those without. Finally, the office of the Surgeon General of the Army, in its May 1875 report made the following recommendations for that year:

> ...the chief cook of each company shall be a permanent detail, and shall receive extra-duty pay that they shall be specially enlisted for that purpose, and that a school for the instruction of cooks be established at the recruiting depots.
>
> ...a manual of instructions for Army cooks be prepared and issued by the Subsistence Department, which shall give diet-table and modes of preparing food suited to the various stations, climates, and seasons.

The task of supplying the army with food, as well as preparing it, remained thorny throughout the nineteenth century. In the early 1800s, the army paid civilian contractors to acquire and transport foodstuffs to military installations, but the contractors often failed to deliver according to agreement, and short rations were a common occurrence of military life—a trying situation, but truly critical when military action was in progress. Getting rations to the men was a recurring concern during the War of 1812 and was General Andrew Jackson's biggest problem during the Florida campaign (1818-1819). In 1818, Congress passed laws requiring that the civilian contractors hired to keep the army supplied with food deliver all orders for rations, in bulk, to designated army depots, where the army would take over, providing stricter control of quality, distribution and transportation. Getting enough food to the fighting forces remained a touch-and-go situation until the end of the Civil War, nonetheless.[36]

Canned foods—the dependable home-pantry standby so taken for granted throughout the twentieth century—came into use, slowly, in the early 1800s. They appeared first in France, where the government offered a substantial cash prize to anyone who perfected a method of preserving and "packaging" food so that large quantities of it could be transported safely and easily. What the French government wanted was a better way than they had for providing rations for military forces. A chef named Nicholas Appert won the prize in 1810. His method involved putting the food to be preserved into glass jars, corking the jars tightly, and cooking the food inside them. The process worked, but the glass containers of that day were bulky, expensive and breakable. It wasn't until 1858 that New Englander John. L. Mason perfected and patented the home-canner's Mason jar with its porcelain-lined zinc screw top, sealed with a flat rubber ring. Very quickly after Apper's successful canning process became widely adopted, however, actually before the end of 1810, an Englishman, Peter Durand, patented the first tin container for preserving foods, and the first U.S. patent for a similar container was issued in 1825. Fish and fruit were among the first foods canned in quantity. Around 1815, a New Englander named Ezra Daggett obtained a U.S. patent for canning seafood, pickles, jams, and sauces. By the 1820s, canned foods were being produced commercially in Boston and New York. By the 1840s, grocery shelves carried canned herring, sardines, and oysters (salmon came later), as well as canned tomatoes, corn, peas, and some fruits.[37]

In 1856, Gail Borden, a Texas surveyor, land agent, and inventor, perfected a way of condensing and canning whole milk and obtained English and American patents for the process. And just before the outbreak of the Civil War, in 1861, Gilbert Van Camp, a former Indianapolis tinsmith turned grocer, perfected a way to can pork and beans in tomato sauce. When the Civil War began, the Union Army bought and issued as a field ration, as much of Borden's canned condensed milk as it could get and made substantial contracts with Van Camp for case after case of his pork and beans,[38] later reported to be a mainstay in the diet of General U.S. Grant who lived strictly on soldiers' rations.[39]

Wartime requirements of the armed forces for more food spurred canners to produce more and more canned meats to supplement the canned fruit and vegetables. Chicken, ham, and "corned beef" were soon being shipped to army kitchens (see Figure 4.13) along with the popular tomatoes, corn, peas, and apricots. In "army camps on, on gunboats, and in [field] hospitals," many men of the 1860s got their "first taste of canned foods" from military kitchens. Borden's condensed milk, highly esteemed by the army, was soon being sold internationally, providing bacteria-free cow's milk for babies and hospital patients long before pasteurization was understood. Canned tomatoes became almost a staple in army kitchens and were "worthily conspicuous in the supplementary rations that the U.S. Sanitary Commission sent to military hospitals."[40]

Often, soldiers used company funds to purchase canned food to supplement their diets. The company fund was used to buy a great many things besides food, however. The Surgeon General's report in 1875 contains the following notation:

> The mess-furniture of a company is provided from the company's fund, and some companies have a very good supply, in which they take much pride. It seems to me, however, that the Government should provide plates and knives and forks

as well as pots, blankets and relieve the company from this source of expense.

That recommendation, in part, was accepted in the 1880s.

As the position of cook in the army became established, the company cook was provided with his own quarters, usually a room near the kitchen. This arrangement can be seen clearly in Figure 3.17, the Quartermaster's plan for the first floor of a two-story barracks. Like a good many things in the military life, it would appear that various individual army installations were detailing permanent work as cooks to those they considered qualified for it before there was any real authorization to do so.

The Cook's Creed

Cleanliness is next to godliness, both in persons and kettles. Be ever industrious, then, in scouring your pots. Much elbow grease, a few ashes, and a little water are capital aids to the careful cook. Better wear out your pans with scouring than your stomachs with purging; and it is less dangerous to work your elbows than your comrade's bowels. Dirt and grease betray the poor cook and destroy the poor soldier, while health, content, and good cheer should ever reward him who does his duty and keeps his kettles clean.

In military life, punctuality is not only a duty, but a necessity, and the cook should always endeavor to be exact in time.

Be sparing with sugar and salt, as a deficiency can be better remedied than an overplus.

Remember that beans, badly boiled, kill more than bullets; and fat is more fatal than powder.

In cooking, more than in anything else in this work, always make haste slowly. One hour too much is vastly better than five minutes too little, with rare exceptions. A big fire scorches your soup, burns your face, and crisps your temper. Skim, simmer, and scour are the true secrets of good cooking.

[Captain James Sanderson, *Culinary Hints for the Soldier,* 1862]

Figure 4.1. Company Mess Hall, Fort Logan, Colorado, 1891-1895

The first-floor mess room in Figure 4.1 is typical of most barracks buildings in the 1880s and 1890s. For the recommended location and size of mess halls in barracks buildings, see the 1872 Quartermaster's floor plan in Figure 3.17.

The regulation mess tables shown here are covered with oilcloth tacked down at the edges on all four sides. These metal-legged tables could be folded for transport, as needed in the field. The benches are smaller versions of the table. The ironstone dinnerware might have a regimental and/or company insignia on each piece, since most equipment of this kind was purchased out of regimental and company funds. As soon after meals as they could be washed and dried, plates, cups, and saucers were returned to individual place settings at the table—cups and plates set upside down. Bowls, platters, and other serving pieces were stored on the mess room shelves, seen at right rear.

The small window just above the china storage shelf on the rear wall opens directly into the kitchen. The door at rear leads to a pantry on the left and a cook's room beyond that.

This mess hall is lighted by the army's Model 1881 barracks lamp, suspended from the ceiling at left. The lack of a wood or coal stove is explained by the presence of the hot-air furnace register in the floor at left foreground. Although the mess hall is on the first floor of the barracks building, there are no window shades or curtains.

[Negative Number 92-F-34-7, National Archives, Washington, D.C.]

Figure 4.1

Figure 4.2. Company Mess Room, Fort Snelling, Minnesota, ca. 1893-1895

Possibly the most striking feature of the mess room in Figure 4.2 is its highly individualistic stove at center rear—clearly recognizable as non-regulation. Lighting was provided by a pair of oddly mismatched wall-mounted lamps, difficult to see because of the lighting tin the room when the photograph was made. The lamp on the right is the army's Model 1881. The one at left is the army's Model 1891.

Rows of stools provide individual seating at two of the mess tables in Figure 4.2. This type of seating for mess halls made its first appearance in the 1880s.

Of special note in this photograph is the mess table at left rear. Unlike the other three mess tables, which are a standard model that appeared around 1884 and 1887, the table at left rear is set on sawhorses like the tables in Figure 4.8. I believe that these were earlier types of mess tables. Resting the tabletop on sawhorses may have produced the army's first collapsible tables. Use of tables that could be quickly taken apart would make it possible to transform a company dining room in a very short while into a reception hall that could be used for a variety of purposes. Much army furniture was collapsible. The other three mess tables shown in this photograph are the new folding tables.

[Negative Number 92-F-65-8, National Archives, Washington, D.C.]

Figure 4.3. Company Mess Room, Fort D.A Russell, Wyoming, ca. 1892-1895

The photographer who shot the scene in Figure 4.3 preserved one corner of the 1890s mess room at Fort D.A. Russell, in party regalia. The customary oilcloth table covers have either been replaced by, or covered over with, what look like lace cloths. Fresh flowers, fruit, and a punch bowl indicate that this is indeed a festive occasion. A handsome Model 1891 barracks lamp hangs above the decorated tables, and the Model 1884 barracks stools are drawn up to provide individual seating.

[William Henry Collection, St. Louis, Missouri]

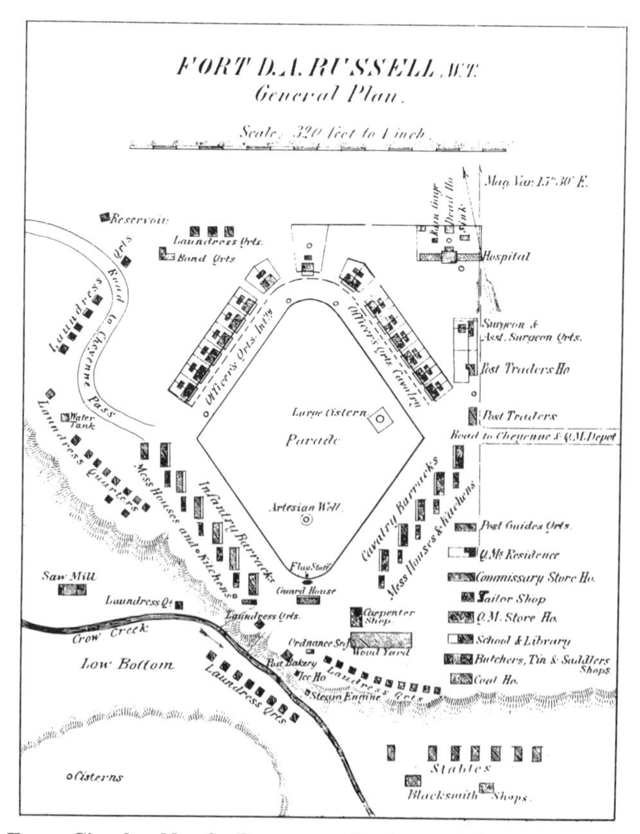

From Circular No. 8, *Report on Hygiene of the U.S. Army.*

Figure 4.2

Figure 4.3

Figure 4.4. Company Mess Room, Fort Sill, Oklahoma, 1897

Clearly, the Fort Sill mess room in Figure 4.4 is being prepared for a holiday meal. The mess room crew has been uniquely inventive with the materials at hand. The centerpieces appear to be fresh celery stalks and bread sticks. Individual place settings are made festive—and colorful—with large apples displayed atop bowls turned upside down to make pedestals for them. Servings of sliced peaches at each place are also made part of the decorations, placed in shallow bowls on top of inverted dinner plates.

Light for the party will be provided by the army's Model 1891 barracks lamp at upper right, and the room will be heated by the wood stove (sharp right foreground)—the same type as the stove shown in Figure 3.27 at Fort McIntosh, Texas, with the bold ''U.S.'' insignia on its sides and back. The blanket at the window is not a sober military window treatment, it was put there by the photographer to control the lighting while the photograph was made.

[Negative Number RG-222-SC-105632, National Archives, Washington, D.C.]

Figure 4.5. Company Mess Room, Fort Monroe, Virginia, ca. 1895

The Fort Monroe mess room in Figure 4.5 is shown decorated for Christmas dinner. Cuttings of holly boughs and ropes of Christmas greens make the decorative electric light fixture at upper right center the ornamental focus of this section of the room. Wiring for use of electricity in army buildings made its way in gradually, but had been in use for some time, in some areas, by the 1890s.

Another innovation that made life easier for the army's cooks and mess crews was the ice box for food storage. Ice boxes of that time were usually made of wood, heavily insulated on the inside, and what looks like an early example of a fine one appears against the rear wall, at left.

The mess room tables—standard army collapsible types—are made festive with white cloth coverings and what appear to be fresh celery-stalk centerpieces. Individual loaves of freshly baked bread are at each diner's place. Diners are to be seated on the army's old-fashioned wooden plank benches instead of on individual stools or folding benches.

A look at the number of dishes already on the tables and the reserve stacks on the storage shelves at right indicate that the Fort Monroe company funds put quite a tidy amount into the purchase of ironstone.

The original photograph made of the scene in the 1890s was used to produce the picture that appears in this book. Still in excellent condition, the original bears, in the lower right corner, the photographer's imprint: ''Cheyne, Hampton, Va.''

[Robert Borrell Collection, Washington, D.C.]

Figure 4.4

Figure 4.5

Figure 4.6. Regimental Mess Hall, Near Washington, D.C., 1864-1865

The photograph in Figure 4.6 was one of a pair made for stereoscopic viewing, a popular Victorian parlor entertainment of the late nineteenth century. The original photograph took in a great deal—which makes it difficult to distinguish details about objects at the far end of the picture. A magnifying glass helps the viewer to see that the bushy, irregular, dark shapes seen at intervals above the tabletops along the tables' length are small Christmas trees, decorated to help celebrate the fort's Christmas dinner. Each tiny tree stands in a large, flat dish filled with apples.

With a very *strong* magnifying glass, the viewer can see, standing at the back of the room, at the end of the tables, some six or seven women in long party dresses. These ladies could be officers' wives who helped decorate and prepare the tables.

The tables and benches are very early types: simple planking nailed together. The building is a temporary one with what appears to be a dirt floor. The mess plates and cups are tin, but the decorations are splendid creations.

The sharp-eyed viewer may note that the table at left clearly has a formal cloth covering, while it is difficult to tell about the one at right. One might be moved to the ungenerous suspicion that the table at right has nothing more than the usual everyday oilcloth.

[Negative Number 111-B4069, from a stereoscopic pair, National Archives, Washington, D.C.]

Figure 4.6

Figure 4.7, 4.8, and 4.9. Consolidated Mess Halls at Three Barracks, 1892-1897: Jefferson Barracks, Missouri; David's Island, New York; and Fort Riley, Kansas

The mess halls in Figures 4.7, 4.8, and 4.9 are unusual in that each is a very large, consolidated mess hall, and few nineteenth-century U.S. Army posts had such extensive facilities. At the few posts where they did exist—large regional training centers—these huge mess halls would remain in use well into the twentieth century.

The hall shown in Figure 4.7 served the Cavalry Recruit School at Jefferson Barracks. The one in Figure 4.8 served the David's Island Infantry Recruit Training Post. These two buildings look very much alike, and so they should; they were built from the same blueprint.

The three photographs in this grouping provide a nice overview of the varied types of furniture available for use in such facilities. In Figures 4.7 and 4.9, the tables are solid, non-collapsible items with sturdy, solid wooden legs firmly attached. The tables in Figure 4.8 are the collapsible type, with the tops resting on sawhorses.

For seating, Figure 4.7 is the aristocrat of the three. It has individual wooden chairs, with backs for diners to lean against. Figure 4.8, with the collapsible tables has the old-fashioned solid, non-collapsible wooden benches. Figure 4.9 has stools for everybody; it also has a marvelous array of the army's handsome Model 1891 barracks lamps and innovative serve-yourself coffee urns at each table.

[Negative Numbers 92-F-30-8, 92-F-19-11, and RG-111-SC-107693, National Archives, Washington, D.C.]

Figure 4.7

Figure 4.8

Figure 4.9

Figure 4.10. Banquet, Headquarters Building, Fort Marcy, Santa Fe, New Mexico, 1885-1893

In Figure 4.10 we see a regulation mess table set up in Fort Marcy's headquarters building for a special celebration, perhaps the Fourth of July, judging by the extensive use of the flag as a decorative motif. Information about this photograph does not include any explanation for the variance in dress of those present. Some are in civilian clothes, others in uniform. It may be that the men in civilian clothes are distinguished local visitors—possibly civil officials from Santa Fe—or it could be that they are non-commissioned officers who chose to appear in civilian dress.

[Negative Number 1706, Museum of New Mexico, Santa Fe, New Mexico]

Figure 4.10

Figure 4.11. Banquet, Regimental Mess Room, 10th U.S. Cavalry, ca. 1895

Background information for the photograph in Figure 4.11 tells us that this formal mess has been prepared for the officers of the 10th U.S. Cavalry. The table is set for thirty-six men, and the decorations make it clear that the event to be celebrated is an important one apparently involving a patriotic theme (indicated by the huge garrison flag at the base of the small stage at rear) and the achievements of the regiment (indicated by the array of company guideons—regimental identification flags—arched above the national flag). A pennant showing the regimental colors appears at the center of the guidon arrangement. The regimental punch bowl appears on the sideboard at left.

Judging by the musical instruments visible on the stage behind the flags, the regimental bandsmen who are able to provide musical entertainment are already taking their places.

[John Bigelow Papers, Library, U.S. Military Academy, West Point, New York]

Figure 4.12. Kitchen, Company Mess, 1st Cavalry, 1890

Figure 4.12 shows the standard arrangement for a company mess kitchen of the nineteenth-century U.S. Army. This rare view shows the regulation army cookstove quite well. This one is a wood-burner, and the second man from the right is about to put another stove wood stick into the firebox. The stove's "U.S." insignia can be seen on the stovepipe rising from the top of the stove to meet the flue pipe.

The older man at right, who is less interested in what the photographer is doing than are his younger helpers, is probably the cook. The other three are probably the current mess stewards, enlisted men assigned to the cook as kitchen helpers on a daily or weekly basis.

Kitchen furnishings and equipment include the army's Model 1881 bracket lamp with shade, on the wall at upper right and the big hot water heater behind the stove at left. A row of white items, a bit difficult to identify at first glance, appear on the stove's metal shelf, to the left of the stovepipe. A close look at the photograph's file reveals them to be eight ironstone gravy boats, neatly stacked, the four on the top row resting on the four below.

[Negative Number RG-111-SC-104107, Major Grow Collection, National Archives, Washington, D.C.]

Figure 4.11

Figure 4.12

Figure 4.13

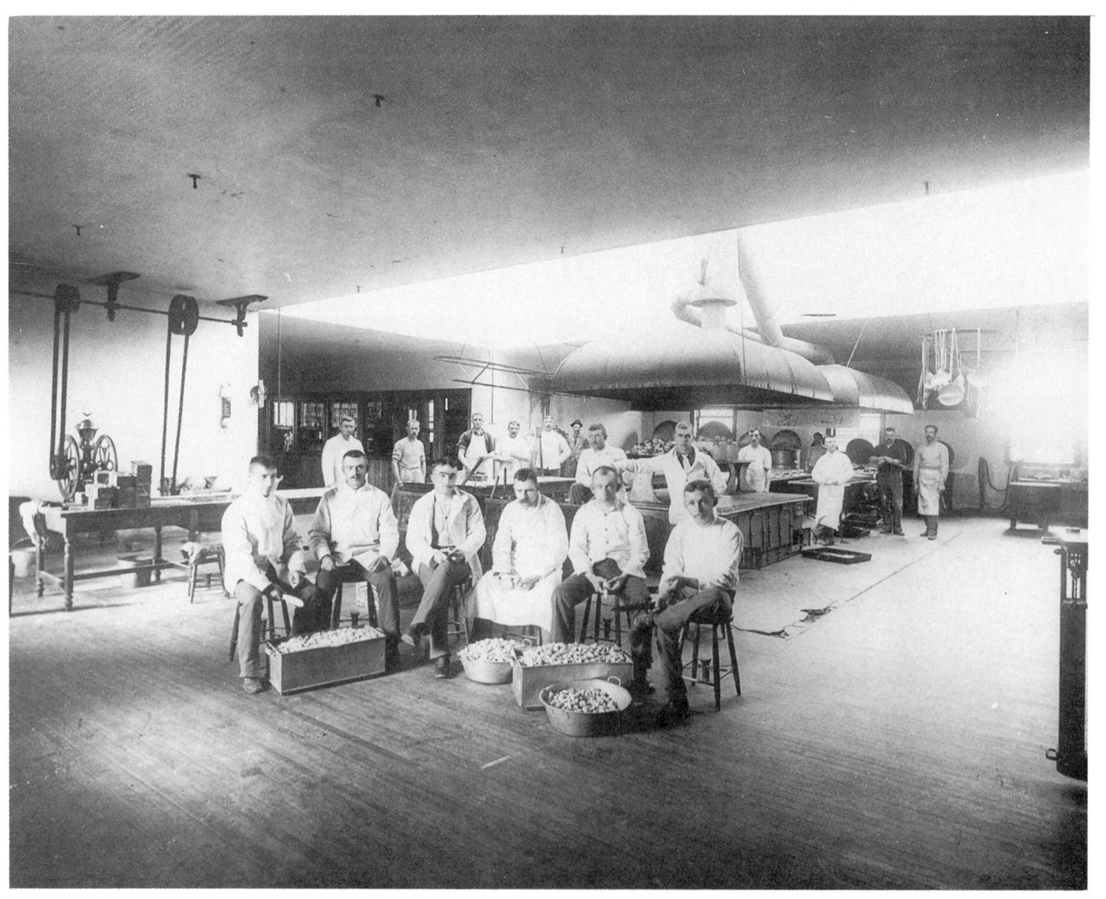

Figure 4.14

Figure 4.13. Kitchen, Company Mess, C Troop, 1st U.S. Cavalry, San Carlos, Arizona, 1897

Background information for the photograph in Figure 4.13 is a bit sketchy, so we do not know which of the four men pictured is the cook, why the second man from the right looks more like a Victorian barkeeper than a soldier on "KP" (Kitchen Police) duty, or why the other man in civilian dress (seated, second from the left) is visiting C Troop's kitchen with a youngster. Perhaps the child, who looks much less hardy than "army brats" were generally assumed to, requires a special diet that must be arranged with the company cook.

We do know that the photograph was made in 1897, and that it is the first photograph in this book to show canned foods on the shelves, available for use in an army kitchen. That made for a great deal more variety in soldiers' meals—a welcome change. For many years, there was so little variety in the menus prepared in army kitchens that, depending on the post's location and the leniency of the commanding officer, the men often cultivated fresh vegetables in a company garden in their off-duty time. At some posts, that activity had become so extensive that inspecting officers grumbled that the men had become nothing but farmers.

The butcher block (right foreground) here in use appears to be a model of efficiency, strength, and simplicity. It consists of a single, solid length of tree trunk sawed from a good sized log. With both sawed ends smoothed and leveled and the bark removed from its sides, it makes a very serviceable butcher block.

[Negative Number RG-111-SC-107685, National Archives, Washington, D.C.]

Figure 4.14. Kitchen, Consolidated Mess Hall, Fort Sheridan, Illinois, ca. 1893-1897

The happy fellows on KP in the foreground of Figure 4.14 are engaged in that classic chore that tradition would have us believe soldiers on KP are assigned to in perpetuity, in this life as well as the next one. They are peeling what must eventually seem to be endless tons of potatoes. A kitchen the size of this one could service a mess hall like the army's other consolidated mess halls, pictured in Figures 4.7, 4.8, and 4.9, so perhaps these men do face endless tons of potatoes.

[Negative Number 92-F-61A-18, National Archives, Washington, D.C.]

GARRISON SUPPORT BUILDINGS

A nineteenth-century military post was a self-contained, interdependent, community, much like a university campus or a religious compound. Similarly, military garrisons needed many of the same kinds of ordinary service facilities and structures always required in such places—essential but usually humdrum buildings such as the commissary, needed for food storage, preparation, and service; the quartermasters' stores, where clothing, tools, and equipment could be stored and dispensed; the dispensary and/or hospital for the ill or wounded, ordnance buildings for weapons and ammunition, and chapels, because people everywhere needed them. Most military posts maintained such structures, with the office space each specified unit needed for the ever-present paper work involved.

Support buildings such as those shown in the following section represent the quieter, more day-to-day aspect of military life, certainly, but over the long haul a soldier spent far more time in garrison and in buildings like these than he did in the field, on patrol, or in combat. Army life has its classic adages and one of them sums it up neatly: "99 per cent boredom followed by 1 per cent sheer terror." It is important to see the areas that nurtured the "99 per cent boredom" to bring more clearly into focus our glimpses of the routine, daily life of the nineteenth-century soldier.

Photographs of the interiors of support buildings for nineteenth-century military posts are quite rare, and the grouping that follows is one that we are very pleased to have assembled.

Figure 5.1. Sutler's Store, Fort D.A. Russell, Wyoming, 1893-1897

The word "sutler" does not appear in all dictionaries today; sutler's and their places of business had begun to disappear sometime before the end of the nineteenth century, and they are today, for the most part, in the same category of earlier Americana as the itinerant tinker and the peddler. *Webster's New World Dictionary* (1892) notes "fading away" in its definition of sutler as "formerly, a person following an army to sell food, liquor, etc. to its soldiers." The definition that best fits the Fort D.A. Russell sutler's store in Figure 5.1—and most other sutlers' stores serving the U.S. Army in the 1800s-is in G. & C. Merriman Company's *Webster's Third New International...Unabridged* (1971). Here, a sutler is defined as a "provisioner to an army post, especially when established in a shop on the post."

Throughout the nineteenth century, especially in the earlier half before the army developed its own commissary services, the U.S. War Department licensed independent civilian traders called "sutlers" or "post sutlers" to establish and operate what amounted to small general stores—"sutlers' stores"—on government property or near each U.S. Army post. Although the post sutlers could sell their merchandise to civilians, they were primarily to sell commodities needed by military personnel, and the army regulated both the kinds of goods sutlers could sell and the prices they could charge for them. Sutlers' merchandise was usually the same homey hodgepodge stocked later in rural general stores: tobacco, combs, pocket knives, boot laces, hard candies, sugar, coffee, salt, flour, kerosene lamps and their wicks and chimneys, kerosene, bolts of cloth, needles and thread, long johns, miscellaneous hardware, and other small items that made life easier in isolated areas. The sutler's store in Figure 5.1 seems well stocked. The shelving behind the counter carries packaged units of a great many different things. In the late 1890s, the country was moving into an era of elaborate packaging, putting everything saleable into containers.

The shelves immediately to the right, behind the soldier, a part-time employee, are lined with shelf paper, decoratively edged. The glass case to the soldier's right contains briar pipes. The sprucely dressed civilian at the corner where the counters meet is the sutler, a proud storekeeper.

[William Henry Collection, St. Louis, Missouri]

Figure 5.1

Figure 5.2. Sutler's Store, Fort Keogh, Montana, 1893-1897

The shirt-sleeved man intent on the paper he holds as he leans against the counter at right in Figure 5.2 is the sutler/owner of the 1890s Fort Keogh post sutler's store. The three unidentified men in uniform nearby are either potential customers or drop-in visitors from the post.

A thoughtful viewer's reaction, after a good look at the interior of this store, may very likely be that this sutler perhaps gave more thought to saving money on construction costs than making the place attractive to potential customers. Montana winters can be fearfully cold, but apparently no provision for heating this store's main lobby was made; there is no stove, no fireplace, no heat register, and no chimney for venting if any of them should be installed later. The store's office space at back, however, the enclosed, windowed area, is shut off from the main lobby and could be kept quite comfortable with a small stove and a modest amount of fuel.

The lighting system is meager also, a single fixture just above the sutler's head appears to be the only artificial lighting provided , and frugality clearly won out over the aesthetics in the selection of flooring, which appears to be rough, unfinished lumber.

[Negative Number RG-111-SC-98015, National Archives, Washington, D.C.]

Figure 5.3. Sutler's Store, Fort Dodge, Kansas, May 25, 1867

The Fort Dodge sutler's store of the 1860s in Figure 5.3 was sketched by an artist for *Harper's Weekly* in May 1867, some twenty-five years or so before the more recent scenes in the stores at Fort D.A. Russell and Fort Keogh (Figures 5.1 and 5.2) were photographed. Like other sutler's stores, the one at Fort Dodge catered to civilians as well as to soldiers, and the artist included a colorful array of its customers, among them an Indian woman, shopping as many mothers do today, with her child tucked into a backpack.

To meet the needs of such a broad range of customers, the Fort Dodge sutler kept a wide variety of merchandise in stock. It was sold or traded unpackaged, on open shelves, as was most general merchandise everywhere at that time. The only packaging done was for bulk items—such things as sugar, flour, salt, coffee, tea, seed, dried beans, and peas—things that had to be dipped out from barrels or bins, weighed, priced, and put into some kind of container before the customer could take it home. Packaging as we know it today was nonexistent. The paper bag, the simplest, most convenient kind of packaging, didn't exist until the 1850s, when an Englishman began selling to grocers in his are the paper bags he made by hand. By 1852, a rudimentary machine for making paper bags was put into operation in Bethlehem, Pennsylvania. In 1860, Charles H. Morgan of Philadelphia designed a better machine and opened a paper-bag factory. And in 1867, Luther C. Crowell, a New Englander, designed and patented a still better paper-bag machine, and followed that up a few years later by designing the square bottomed bag so familiar today.[41]

The labeling of items for sale was almost nonexistent early in the nineteenth century. Labels as we know them were used, to some extent, as far back as the late 1850s, but they, too, did not attain significant popularity or widespread use until the 1870s and 1880s.

[Fort Dodge, Kansas, as sketched for *Harper's Weekly,* May 25, 1867. Kansas State Historical Society, Topeka, Kansas]

Figure 5.2

Figure 5.3

Figure 5.4

Figure 5.5

Figure 5.4. Post Trader's Store, Fort Bridger, Wyoming, ca. 1895

The photograph of a post trader's store at Fort Bridger in the mid-1890s (Figure 5.4) is the only photograph I have seen of a civilian-run meat market on a nineteenth-century army post.

Cured pork hams and shoulder and sections of bacon ("middling meat" or "middlings") are conveniently stored and displayed on the store's back wall. Fresh meats could be kept for brief periods in insulated ice boxes cooled with blocks of ice—either natural or the artificial kind which was being produced in the larger cities by the mid-1800s. The electric light bulb fixture (top left foreground and above the rear counter) tells us that this shop was wired for electricity, but electric refrigerators were not perfected and generally available until early in the twentieth century.

This store's general sales room had obviously served a different purpose at some earlier time, judging by the delicate border stenciling on the walls and the large circle visible on the ceiling. Furnishing for use of the place as a market include the wall clock near the mounted antlers at left rear and the sturdy counters, decorated with a three-color paint scheme, a technique quite popular in the late 1800s.

[Wyoming State Archives and Historical Department, Cheyenne, Wyoming]

Figure 5.5. Issue Room, Small Stores, Fort Ringgold, Texas, 1893-1895

Unlike the stores shown in Figures 5.1 through 5.3, the facility in Figure 5.5 is not a sutler's store, but a link in the army supply chain between the post sutler and today's "PX" (Post Exchange). This Fort Ringgold issue room for small stores was an army-operated successor to the sutler system. As mentioned earlier (Figure 5.1), for the most of the nineteenth century, a sutler's store, located either on the post or very near it, served each U.S. Army post. The sutlers who operated these stores sold army-approved general merchandise at army-established prices to both civilians and military personnel. The army-operated outlets and, later, the quartermaster's stores were open to military personnel only.

Many sutlers gradually branched out, as time went on, and installed tables for card games, checkers, chess, dominoes, and billiards in their establishments,. These additions made the sutlers' stores popular gathering places—much like unofficial clubs—for the post's enlisted men.

Near the end of the nineteenth century, the army began setting up its own sales outlets on military posts, as the Surgeon General had recommended in the late 1870s, some twenty years earlier.

At first, in the late 1880s and early 1890s, the army's stores on post sold only foodstuffs. Gradually, they began adding to the cheese and crackers and canned sardines many of the small luxuries and minor necessities of general store merchandise previously handled by the sutlers.

In Figure 5.5, crates and barrels of commercial products can be seen through the open door of the warehouse to the rear of the store. The good-sized weighing pans on the two sets of scales tended by the soldiers on duty indicate that the store clerks did much weighing, wrapping, and tying-up of bulk item purchases.

[Negative Number 92-F-95-23-A, National Archives, Washington, D.C.]

Figure 5.6. Issue Room, Small Stores, Fort Sam Houston, San Antonio, Texas, September 27, 1877

Figure 5.6 shows the Issue Room for Small Stores at Fort Sam Houston, photographed while the fort was still under construction.

At that time (1877) nearly everything was visible on the open shelves seems to have been stored in bottles or jars. The back wall looks like the pharmaceutical section of a present day drug store. The extensive wooden counters provided a great deal of storage space, with additional room in large drawers under the counter tops, and the tops made convenient space for the clerk to wrap items sold. The counter at right, across the end of the room, appears to be made of two tables joined together.

[Negative Number 165-S-284, National Archives, Washington, D.C.]

Figure 5.7. Quartermasters' Supply Room, Fort Ringgold, Texas, 1893-1897

Typical of quartermasters' supply rooms established at major U.S. Army posts in the late nineteenth century is the one at Fort Ringgold (Figure 5.7). These early quartermasters' supply stores gradually replaced the sutlers' stores that had formerly supplied merchandise to post soldiers and civilians. The quartermaster's outlets were open only to military personnel, however. They were a part of the service system provided by the Quartermaster Corps, the army's supplier of quarters, food, clothing, and equipment for army personnel. The Quartermaster Corps remained in operation until 1946, when it was replaced by the U.S. Army Material Command.

A good quartermaster's supply store in the 1890s stocked a number of things that might be needed or wanted on an army post some distance from civilian sales centers. Varieties and amounts of merchandise offered can be seen in goods under and on the open shelves and hanging from the ceiling of Fort Ringgold's supply room.

Sometimes clothing, supplies, and equipment were all handled in the same quartermaster's supply building. Sometimes, as was true at Fort Ringgold, supplies and equipment were available in one building, and uniforms and work clothing were handled in a different building (see Figure 5.8).

The paper work at Fort Ringgold's general supply outlet was done at the sergeant's desk (center). Light was provided by the Model 1891 barracks lamp suspended above the desk, and the Model 1891 barracks bracket-mounted lamp to the right of the desk.

The photograph for Figure 5.7 was probably taken in summer, since the soldiers on duty in the supply room are wearing light summer uniforms that were available in either linen or cotton.

A water cooler (far right foreground) helped the Ringgold quartermaster's staff keep cool. This cooler is clearly not a showpiece, but an object of use. The metal bucket stationed below its spigot was there to catch drips and spills as the spigot was turned on and off. Coolers like this one appear in many photographs of army buildings of this period. They must have been in use everywhere.

Behind the sergeant's desk is an open-platform freight elevator. The machinery used to operate it—a huge wheel and a thick rope belt—was left out in the open, just beyond and to the left of the desk. Such safety precautions as walling off or enclosing the machinery and the elevator shaft, were, at that time, refinements still to come.

A building almost identical to the one in Figure 5.7 can be seen today at Fort Laramie National Historic Site, Wyoming.

[Negative Number 92-F-95-20-A, National Archives, Washington, D.C.]

Figure 5.6

Figure 5.7

Figure 5.8

Figure 5.9

Figure 5.8. Quartermaster Stores: Uniforms, Fort Ringgold, Texas, 1893-1895

Fort Ringgold had two post quartermasters' supply installations in the 1890s. Uniforms and work clothing were issued from a unit (Figure 5.8) separate from the one that carried general supplies and equipment (Figure 5.7). The uniform supply store stocked not only all the ready-made components needed to make up a complete U.S. Army uniform, it also carried bolts of whole cloth, so that those who wanted—and could afford—well tailored clothing could be measured and properly fitted by a company tailor.

The sizable collection of ledgers stacked on the end of the counter (right foreground) are clothing record ledgers. In these books, the measurements and clothing record of each soldier on the post would be kept. Judging by its well-stocked shelves and the number of clothing record ledgers in use, Fort Ringgold's quartermaster's store for uniforms must have been a regimental supplier. A quartermaster's store providing clothing for only a single company of soldiers would use only a single ledger.

Like the open shelves of Fort Ringgold's general quartermaster outlet, the shelves here have been built along every available inch of wall space, leaving room at window openings for only window frames, facings, and glass.

[Negative Number 92-F-95-25-B, National Archives, Washington, D.C.]

Figure 5.9. Quartermasters' Supply: Uniform Storage, Fort Grant, Arizona, 1893-1895

Figure 5.9 shows the quartermasters' supply (and storage) room for uniforms at Fort Grant. The Fort Grant facility looks, and is, quite different from the one at Fort Ringgold (Figure 5.8), although the Fort Grant store no doubt also had a "dispensing area" with a counter and various articles of clothing displayed on open shelves.

This photograph was taken to document, in a nomination for commendation, the hard work of the sergeant who planned and built this storage area. Not only did that quartermaster sergeant design and build an extraordinarily well-organized, well-thought-out facility for that day and age, but he did it using nothing but scrap lumber and packing crates as building materials.

In the dusty Southwest, such protective closed-storage units as these were clearly a great improvement over the usual open-shelf issue rooms of most quartermasters' clothing supply outlets of this era. Marked on the door of each of these units is he type and size range of the clothing inside. Small slips of paper fixed to each door were a running inventory system. When an item of clothing was removed from a compartment, one slip was torn off the paper pad on that compartment door, so that the number of articles left in the unit could be quickly determined without even opening the door. The sliding ladder, also an innovation in the quartermaster supply room at that time, made it easier to reach articles needed from top compartments.

[Negative Number 92-F-24-9, National Archives, Washington, D.C.]

Figure 5.10. Barroom, Fort Riley, Kansas, 1895

As the number of army-operated post exchanges grew, erstwhile post sutlers stopped selling general merchandise and began opening barrooms, pool halls and other off-hours gathering places near army posts. Figure 5.10 shows an enlisted men's barroom at Fort Riley in the 1890s. Clientele was limited to enlisted men; officers did not patronize the bars, lounges or taverns where the enlisted men gathered, and neither did the army wives who lived on the post.

This photograph was apparently made in the summer, since some of the men are wearing sun helmets. The backless stools at the tables are similar to the seating often used in contemporary army mess halls. The barroom floor seems to be covered with sawdust, a common practice of that time. The several large spittoons in evidence, made of either earthenware or treated fiberboard, indicate that the Fort Riley bar customers preferred chewing tobacco or dipping snuff to smoking pipes, cigars, or cigarettes.

[Kansas State Historical Society, Topeka, Kansas]

Figure 5.11. Barroom, Unidentified Military Post, N.D.

Figure 5.11 shows a varied group of off-duty enlisted men socializing at the bar, though they are still in uniform. The "three-looped" sleeve insignia on the uniform of the man seated nearest the bartender (at right) identifies that wearer as a musician. A lone black soldier leans comfortably against the wall at back, and the mustached man happily posing for the camera (front and center) is either a civilian or a soldier who doffed his uniform for "civies" before dropping at the bar.

The paintings decorating the walls here hang unusually high—until one reflects that, very likely, groups of men often leaned against these walls, and pictures hung any lower than they are here would soon be crushed or knocked to the floor. The shining spittoons visible here are similar to those in Figure 5.10, and, like them, appear to be made of either earthenware or treated fiberboard.

Safety measures to protect the barkeeper—and the till—were built in when the bar was installed. A hinged panel with a quick-release mechanism is secured above the barkeeper's head. In case of trouble, the barkeeper can release the mechanism and the panel will drop instantly, covering the bar window and closing off the barman and the area behind the bar from the rest of the room.

[Negative Number American West 63, National Archives, Washington, D.C.]

Figure 5.10

Figure 5.11

Figure 5.12

Figure 5.13

Figure 5.12. Office, Fort Ringgold, Texas, 1893-1897

The business office pictured in Figure 5.12 is probably a regimental office, rather than a company office, judging by the size and amount of furniture it has and the two soldiers on duty there. A company office would be smaller with fewer furnishings, and one individual would handle the paper work.

The army has furnished the Fort Ringgold office well. The desks are of impressive size, well-constructed, and well-kept. An excellent view of the army's Model 1891 barracks lamp almost diverts the viewer's attention from the unusual piece of furniture against the wall beyond the lamp—a very tall, well-built, well-stocked bookcase with glass doors to protect the books stored there. An early model typewriter is visible, next to the large water cooler at right, and several metal spindles or spikes, for quick temporary storage of miscellaneous papers, appear on the desk tops. What appear to be dark pull shades hang at both windows, and the lower part of the large "roller" wall map can be seen on the wall at right, just above the water cooler.

[Negative Number 92-F-95-21A, National Archives, Washington, D.C.]

Figure 5.13. Senior Officer's Station, Fort Sam Houston, San Antonio, Texas, 1893-1897

The office photographed in Figure 5.13 is probably that of the post adjutant, judging from its location outside the commanding officer's suite, and the bearded man at the desk is probably the adjutant. The post adjutant served as an aide to the commanding officer of an army post and, in most situations, acted as chief administrative officer for the post. Most of the paper work involved in the day-to-day business of running the post was in the hands of the adjutant.

The senior-officer status of post adjutant is reflected in his office furnishings: the substantial desk and its cane-backed chair—probably a swivel model matching the one seen through the doorway in the post commander's office—the revolving bookstand at left, and, at right, the capacious closed bookcase or wardrobe of burnished wood, probably used to store files, ledgers, and supplies, and the carpeting, which appears to be Brussels. The carpeting in the commander's office may be ingrain. The big roll-up map of North America on the wall behind the seated officer appears to have been well used.

[Negative Number RG-111-SC-87370, National Archives, Washington, D.C.]

Figure 5.14. Rosebud Indian Agency Office, Sketched by James McCoy, May 3, 1879

The "Rosebud Agency Office" shown in Figure 5.14, as sketched by James McCoy in May of 1879, was the building in which agents appointed by the federal government's Bureau of Indian Affairs administered governmental programs designed for the Indians living on the Rosebud Indian Reservation in western South Dakota. The agency office is included in this section of the book, among army post support facilities because although it wasn't built on the army post, the Indian Agency building was a government-ordered structure, a part of the federal government's frontier system for establishing and maintaining law and order. Since the only real regulatory powers of enforcement that the Indian agents had access to were the soldiers at the nearest U.S. Army post, all Indian agents kept in very close touch with the post's soldiers and commanding officer.

The Rosebud Reservation administered by the Rosebud Indian Agency was one of six Teton Sioux reservations established by the Dakota Territory in the late 1870s. Farthest westward, toward Wyoming, bounded on the south by what is today the state line between South Dakota and Nebraska, was the Pine Ridge Reservation. The Rosebud adjoined it, immediately to the east, the boundaries between them indistinguishable. Just north of them were the Standing Rock and Cheyenne River Reservations, with part of Standing Rock in North Dakota. In between lay a third—and considerably smaller—pairing, the Lower Brule and the Crow Creek Reservations, occupying a modest space on either side of the loop in the Missouri River just below Pierre, South Dakota, today.[42]

Somewhat surprisingly, the word "Rosebud" recurs as a place name more than once in the homeland of the Plains Indians and its history. Near the Custer Battlefield National Monument in Montana is a modest range of hills called the Rosebud Mountains. Looking westward on the map from today's southeastern Montana, in what has been called the heart of the Plains Indian country, one sees the almost mythical names of the rivers flowing into Yellowstone there: the Powder, the Tongue, the Bighorn, and the Rosebud. Early in June of 1876, about thirteen hundred Sioux and Cheyenne Warriors and about the same number of U.S. Army regulars, irregulars and scouts under General George Crook, fought the day-long Battle of the Rosebud along that river's banks. At the day's end, both sides retreated, both claiming victory.[43]

Figure 5.14

In the artist's drawing of the Rosebud Indian Agency office, one agent (right foreground) seated at a handsome desk consults a ledger. Beyond him, toward the rear of the room and turned away from his desk to enjoy a moment's idleness, a colleague rests in a swivel chair. Another agent, standing left center, works atop a long section of cabinets with drawers. A fourth man (background, center) talks through the open half of a Dutch door with a group of people standing outside the building.

The office stove, looking a bit small in comparison to the rest of the furnishings, stands at dead center of the room. Immediately beyond it to the left, against the back wall, is the agency's huge safe which appears in the sketch to extend forward into the room almost as far out as the stove. The three windows are plain, uncurtained and unshaded. The wooden desks and the stand up counter are unusually attractive pieces, possibly constructed on the reservation, itself.

[Negative Number 989-RG-75, National Archives, Washington, D.C.]

Figure 5.15. Dr. Williams' Office, Fort Jefferson, Florida, 1893-1895

Throughout the 1800s, the medical officer assigned to an outlying army post was, in all likelihood, the only doctor in the area for miles around. Because of the scarcity of trained medical help, the nineteenth-century army post surgeon often treated civilians who lived near the post, in addition to treating military patients. All fees collected for medical help to civilians went into the post's hospital fund.

Straw matting covers the floor in this large, well-ordered office waiting room. The two rocking chairs—one of wicker—look as if they would be more comfortable than most waiting room furniture today. A bulky, light-colored fabric—possibly monk's cloth—screens the changing area in the corner, near the window, and a scarf with a cheerful printed border decorates the mantel above the boarded-up fireplace. Artificial lighting, as needed, would come from the student lamp just barely in the picture at far right.

[Negative Number 92-F-29-8-A, National Archives, Washington, D.C.]

Figure 5.15

Figure 5.16. Hospital Ward, Unidentified Army Post, ca. 1893-1897

Army post hospital wards of the 1890s were set up according to the recommendations of the Surgeon General of the Army. Each ward was to have twelve beds, five rocking chairs, a clock, a barracks stove, and adequate lighting.

Model 1893 army bunks are the beds used in the ward shown in Figure 5.16. The clock, stove, and four of the five rockers are visible. Light is provided by the army's Model 1881 barracks lamp, difficult to see, suspended from the ceiling just beyond the stovepipe. Roller shades hang at the windows. The book shelves at the far end of the ward seem well stocked, and next to them, a cabinet with glass doors stores supplies. The photo file indicates that the metal spittoon (center foreground) had a porcelain surface.

[Photograph Number 48, William Taylor Collection, Western History Collections, University of Oklahoma Library, Tulsa, Oklahoma]

Figure 5.17. Hospital Ward, Fort Lawton, Washington, ca. 1893-1897

Fort Lawton's hospital ward (Figure 5.17) is a model army hospital ward for its time, the 1890s. It looks scrupulously clean. The floor —apparently recently repainted—is dustless and shining, the beds all neatly made. Tables between each pair of beds provide space for small items in daily use. Steam radiators provide reliable, even heating. And all the pull shades in the room have been very carefully aligned at the same level, except for the two at the back, which have been drawn, to control the lighting for the photograph.

[Negative Number 92-F-94-15, National Archives, Washington, D.C.]

WAR DEPARTMENT,
QUARTERMASTER GENERAL'S OFFICE.

Specifications for Pillow-cases.

To conform in all respects to the sealed standard sample.

Material.—To be made of fine quality unbleached muslin.

Dimensions.—To be thirty-six (36) inches long and eighteen and one-half (18½) inches wide, when finished.

Workmanship.—To be made in a neat and substantial manner, and to have a two (2) inch seam at the top.

Adopted October 16, 1886, *in lieu of specifications of March* 28, 1885, *which are hereby canceled.*

S. B. HOLABIRD.
Quartermaster General, U. S. A.

1286—F., 1886.

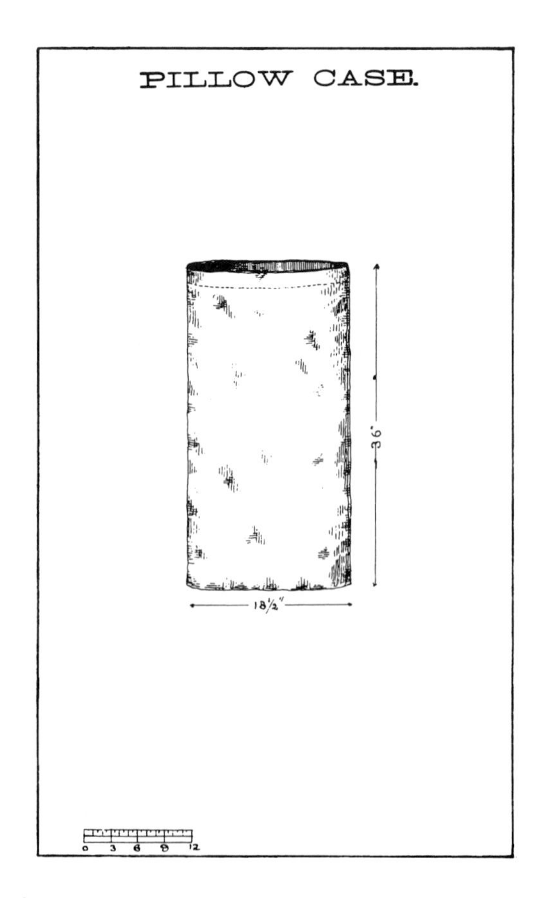

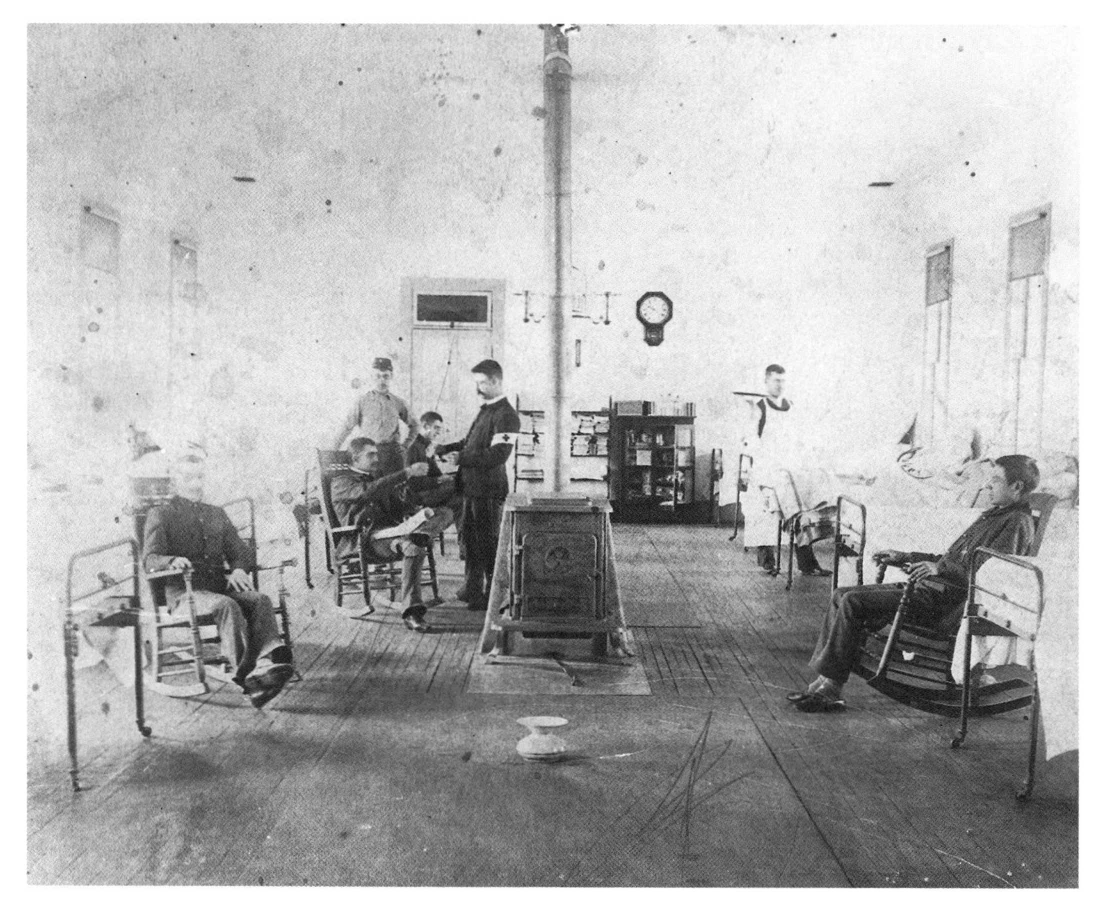

Figure 5.16

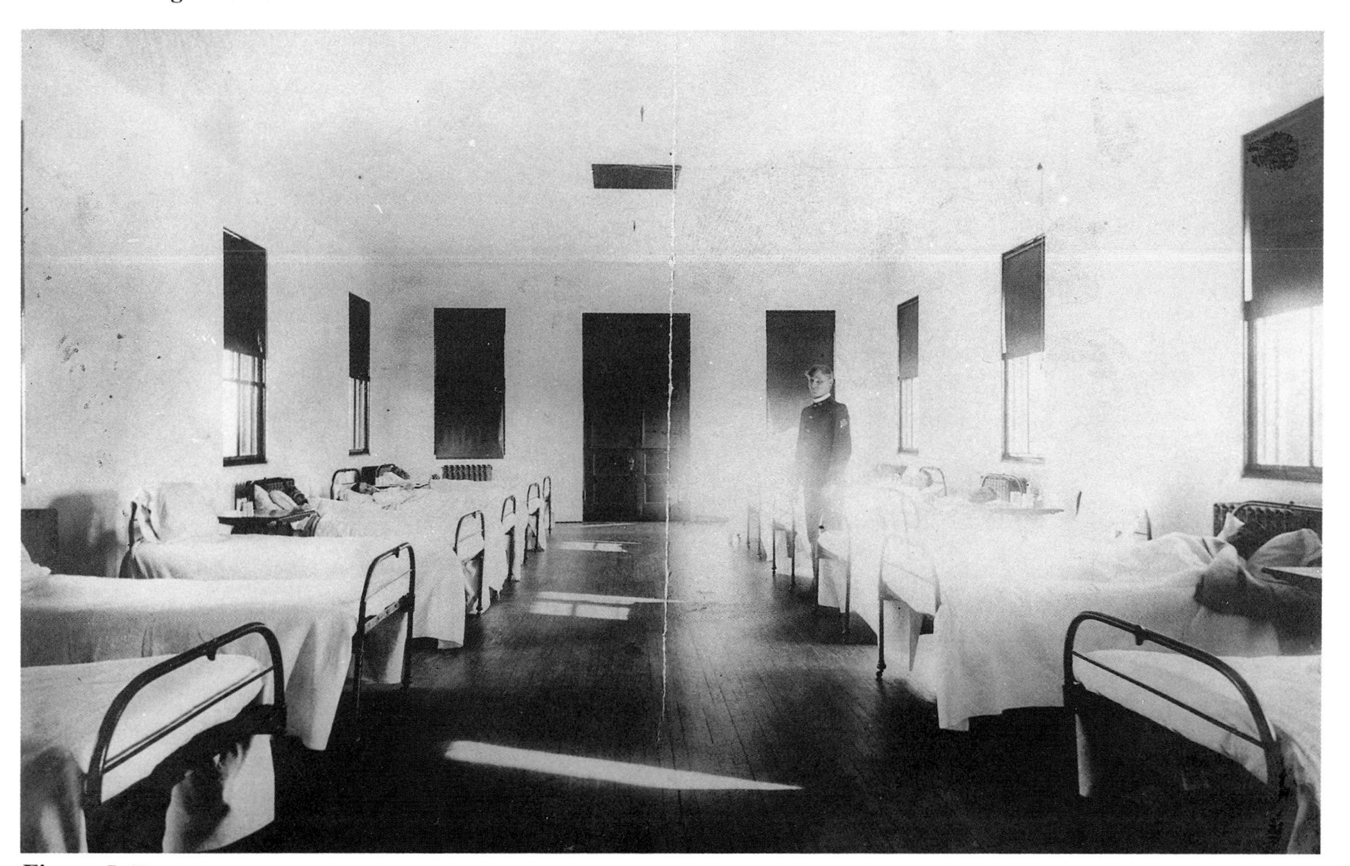

Figure 5.17

Figure 5.18

Figure 5.19

Figure 5.18. Post Chapel, Fort Huachuca, Arizona, 1898.

Judging by the student desks pushed in behind the chairs assembled to sea the congregation during religious services, Fort Huachuca's post chapel (Figure 5.18) also serves as a schoolroom for the children on the post. Two Model 1891 army barracks lamps provide lighting as needed. Taking into account the biblical verse spelled out in the greenery on the back wall, "Glory To God In The Highest," and the amount of fresh green boughs and branches brought inside for ornamentation, I should guess that the scene photographed is decorated for a holiday service, most likely Christmas Eve or Easter.

[Negative Number 93621, Museum of New Mexico, Santa Fe, New Mexico]

Figure 5.19. Post Chapel, Fort Douglas, Utah, 1883

Fort Douglas, Utah, in the early 1880s maintained its chapel (Figure 5.19) for religious services only. Regular church pews provide congregational seating, with a few chairs at pew ends for overflow, and three rows of chairs placed near the chancel to seat the choir. The building appears to have been heated by two stoves, one at either side of the sanctuary. The light fixture appears to be the army's Model 1881 barracks bracket lamp. The inscription of verses of scripture in decorative lettering was often seen on the inside walls of church buildings in Victorian times.

[Negative Number RG-111-SC-88080, National Archives, Washington, D.C.]

WAR DEPARTMENT,
QUARTERMASTER GENERAL'S OFFICE.

Specifications for Pillow Sacks.

Material.—To be made of cotton or linen drilling, or seven (7) ounce cotton duck of good quality.

Dimensions.—Length, when filled, twenty-seven and one-half (27½) inches; width, when filled, seventeen (17) inches; depth, when filled, three and three-fourths (3¾) to four (4) inches. Measurements to be made from corner to corner.

To have an opening or fly in the seam in the upper side seven (7) inches long, to be fastened with two (2) strings of three-quarter (¾) inch cotton tape. Ends of opening to be properly stayed with button-hole stitch.

Ends of sack to be cut square.

Adopted March 12, 1879.

M. C. MEIGS,
Quartermaster General,
Bvt. Major General, U. S. A.

337—Q. M. G. O., 1879, Cl. and Eq. supply.

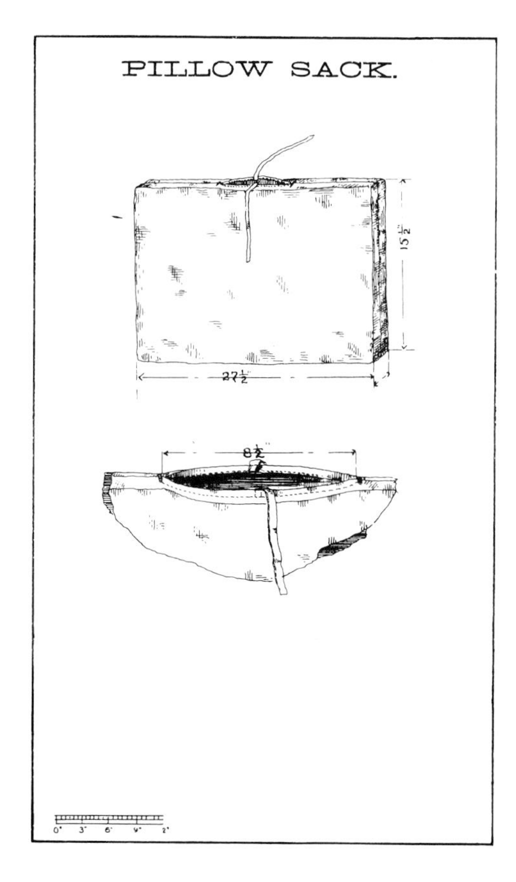

Figure 5.20. Ballroom, Fort Wingate, New Mexico, Christmas Eve, 1896

The men of Troop E, 2nd Cavalry, planned and put up the ballroom decorations (Figure 5.20) for the fort's 1896 Christmas Eve celebration. The huge room is probably Troop E's barracks room, temporarily emptied of everyday furnishings to make it a ballroom for the party. The shelves on the wall between the windows, at left, are ordinary clothing shelves, temporarily trimmed with star-spangled red, white, and blue bunting, with crossed sabers and dress-uniform helmets placed on them as appropriate decorations. The same bunting makes a decorative top-of-the-wall border, with diagonal streamers running from corner to corner. The troops rifle rack, is, for this occasion, decorously hidden behind the door at left. Indian rugs at the two rear windows (background, right and center) were probably put there by the photographer to control the lighting while he made the photograph.

One wonders whether the lifelike forms of the trooper and his horse and the convivial Dickensian figure keeping them company in the ballroom are permanent symbols for Troop E, but the file information for this photograph sheds no light of these three remarkable figures.

[Negative Number 39377. Photograph by Ben Wittich, Museum of New Mexico, Sana Fe, New Mexico]

Figure 5.21. Guardroom, Fort Niagara, New York, 1947

The 1947 photograph of the Fort Niagara guardroom (Figure 5.21) is included here to show a *reconstructed* banquette or "sleeping rack" of the sort built into all British, French, Spanish, and American army guardrooms until the mid-1870s and early 1880s.

These Spartan shelves, modeled after the narrow, raised walkway along the inside of the parapet or protective inner wall of frontier blockhouses, were installed as rough resting places for soldiers standing guard in the cramped quarters of the guardroom. Banquettes were found only in the guardrooms of military installations; they existed nowhere else but on the post. An army post guardroom was traditionally austere. Its only furnishings were the banquette, a gun rack, and a table with benches, none of them intended to encourage lounging.

Assignment to guard duty was for a period of twenty-four hours, during which the guardsman was on duty for two hours, then off duty for four hours, with the cycle repeating until the twenty-four hour assignment was completed. During off-duty hours, the guardsman could rest, or sleep, if possible, on the banquette with his clothes and equipment at hand. No personal effects were ever brought into or kept in the guardroom. Once a soldier's twenty-four hour shift at standing guard was completed, that soldier might not be called up again for a week or even a month.

In addition to the guardroom, the post guardhouse usually contained a small office for the officer of the guard and the post's jail or prison. The jail was one large cell, or room, with smaller cells leading off the main holding area. None of these cells contained bunks or cots. Army prisoners were allowed a blanket and a bucket, and sometimes, if they were lucky, there might be straw on the floor.

Banquettes were not done away with in the U.S. Army until the single iron cot was fully in use at all posts. One aging banquette was still being used at Fort Laramie, Wyoming in 1879.

[Old Fort Niagara Association, Youngstown, New York]

Figure 5.20

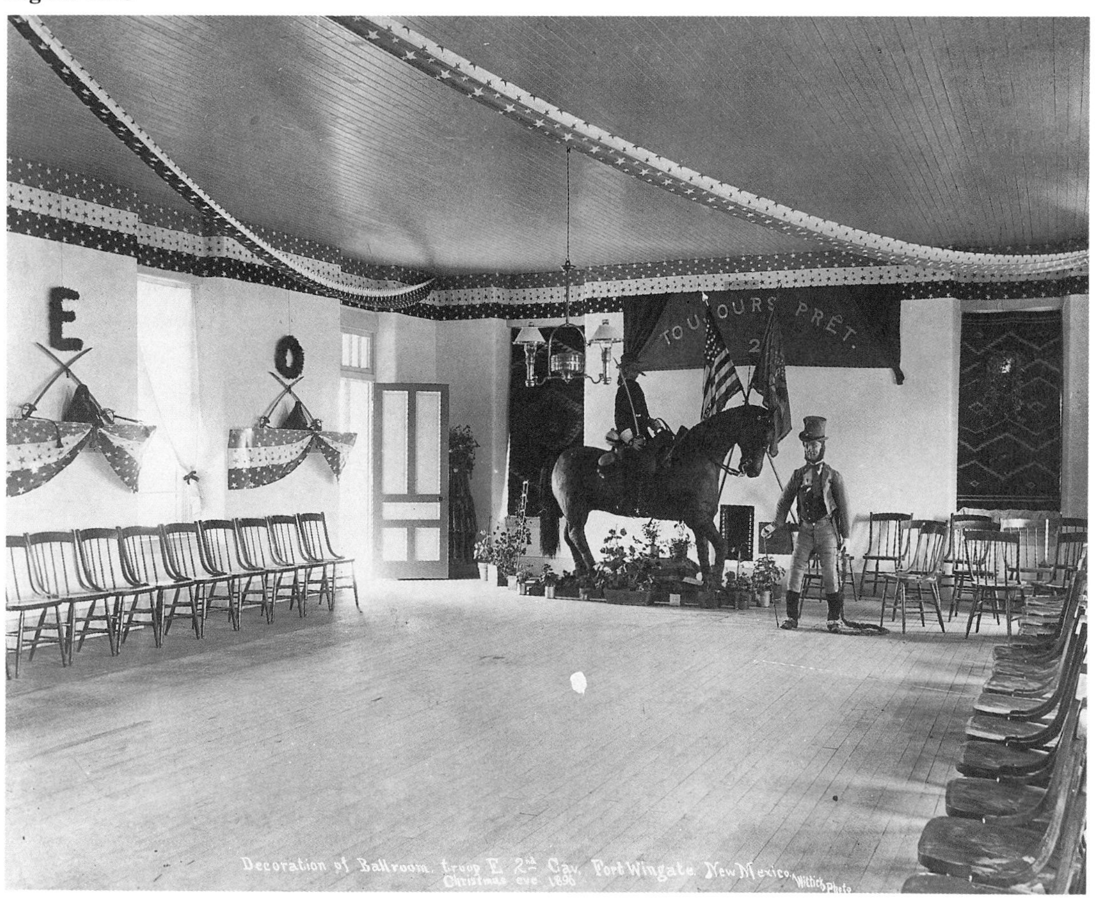

Figure 5.21

Figure 5.22

Figure 5.23

Figure 5.22. Post Library, Fort Crook, Omaha, Nebraska, 1903-1905

All U.S. Army posts had libraries. Throughout the nineteenth century, these post libraries continued to grow, and at the century's turning, many of them had increased their holdings from a few hundred volumes to thousands. In addition to its books, the post library provided a great many newspapers and periodicals, for which the company in residence paid the subscription fees. In 1871, one post, manned by four officers and thirty-six enlisted men, spent fifty dollars a year on newspapers and periodicals; a very considerable sum in those days when an army private's pay was thirteen dollars a month. Reading was a major pastime for army men and their families, however, and it is interesting to note that, by 1900, Nebraska—where Fort Crook was located—was one of those states with the highest rate of literacy in the nation. The other two were Kansas and Iowa.[44]

At all army posts, the holdings of the post library were cherished an well cared for. The Fort Crook Library provided glass doors for the shelves along the one wall photographed to protect what appear to be sets of reference works and other expensive volumes.

[William Henry Collection, St. Louis, Missouri. Photograph by Lester Davis]

Figure 5.23. Military Clerk's Stand-Up Desk, Fort Scott Historic Site, Fort Scott, Kansas, ca. 1830-1850

Figure 5.23 is a present-day photograph of an original piece of military office furniture dating from the mid-1840s—a military clerk's stand-up desk. I include it here simply because this particular item of military office furnishings is so rare. It is put together in much the same way as the military crib of the same period (see Figures 3.1, 3.2, 3.3, and 3.4). Like the crib that served as a model for those photographs, the desk shown in Figure 5.23 also was found in the Philadelphia area.

When the desk was first found, its writing surfaces showed no trace of the cover fabric routinely put on these pieces of furniture when they were built. But when the desk's moldings were removed to replace the missing fabric and make the restoration historically accurate, scraps of blue kersey—blue wool like that used for enlisted men's uniforms of the period—were found, still in place, where the cloth had been protected from wear.

This type of desk would have originally been used in a headquarters building or in any large military office since it could accommodate four clerks at a time, two on either side.

[Fort Scott, Kansas. Photograph by the author]

CADET ROOMS: UNITED STATES MILITARY ACADEMY AT WEST POINT

The United States Military Academy, formally opened at West Point, New York, on July 4, 1802, was not the country's first military establishment on the Hudson River promontory called West Point. Early on, the Hudson was recognized as a vital transportation waterway. To guard it from takeover by the British during the Revolutionary War, Continental troops in 1778 built and occupied a series of forts and redoubts on the promontory's broad, level plain. Rising on the river's high west bank, above a deep, narrow gorge, the West Point promontory was, for the Continental Army, a strategic military vantage point.[45]

The academy's more immediate predecessor at West Point was a small school, authorized by President John Adams in 1801, for training artillery cadets and army gunners and sappers in mathematics, artillery, and fortification.[46] Succeeding Adams as President in 1801, Thomas Jefferson saw greater possibilities for technical training at the small West Point school and ordered it made into a more formal, permanent academy. The following year, Congress established the United States Military Academy there as "a small part of the new branch of the army—the Corps of Engineers."[47]

Creation of the Corps of Engineers and a military academy where engineering could be taught came just at the right time for the country. The Louisiana Purchase of 1803 doubled the nation's size, and West Point historians Dave Richard Palmer and James W. Striker note that "the Army provided expeditionary forces to explore and take possession of the new territory."[48]

By 1853, with other large tracts of land acquired in rapid succession, the nation's original modest size had been increased by some two and a quarter million square miles of new territory.[49] All of it needed to be explored, evaluated, mapped, and surveyed. From the early 1800s until the early 1860s, the country had a pressing need for civil engineers to make reconnaissance surveys, mark territorial boundaries, survey coastlines and inland bodies of water, plan for essential road and canal systems, and, eventually, to plan routes for and build railroads. The Military Academy at West Point was a major source of civil engineers. Historian Nelson Manfred Blake says of it, in *A Short History of American Life:*

> The rapid growth of the country and the building of turnpikes, canals, railroads, and factories demanded trained engineers, but the supply was small. For the first few decades of the nineteenth century, the Military Academy at West Point...was the only institution where such training could be had.[50]

Authors Palmer and Stryker make a similar observation:

> Exploration, surveying, and engineering skills were...part of the soldier's contribution to his nation's development... As more...territory was opened for settlement, the nation exploited its prime source of professional engineering skill, the Army, to aid in developing its transportation network... The Army, through its West Point Graduates, was the only source of the required numbers of trained

civil engineers in the 1820s and 1830s.[51]

Despite its contributions to the country through the training of its graduates, the United States Military Academy had a difficult time of it in the early years. There were shortages everywhere; shortages of buildings, furnishings, equipment, and students, of staff and administrative personnel and, chronically, shortages of funds.

In 1812, Congress increased the size of the Corps of Engineers and expanded the academy's facilities. In 1816, Congress provided money for "buildings, books, maps, and instruments... The facilities and staff...were expanded, the curriculum was broadened, requirements for admission improved, and cadet-gray uniforms were issued for the first time, that fall.[52]

In 1817, Major Sylvanus Thayer, educator, engineer and himself a West Point graduate (1808), was appointed superintendent of the academy. Major Thayer thoroughly reorganized and greatly improved the institution's curriculum, instructional approach and standards of proficiency. During his administration (1817-1833), West Point came to be a source of army officers trained for military leadership instead of primarily a school for training army engineers. The academy was no longer formally related to the Corps of Engineers after 1866. It operates today under the Department of the Army.[53] In the early 1930s, the curriculum was again thoroughly reorganized to establish a strong academic program leading to the degree of bachelor of science and continuing basic military instruction and rigorous physical training.[54]

Today, the West Point military reservation consists of some 15,000 acres, including Constitution Island which faces the promontory of West Point from the opposite bank of the Hudson River. The academy's gray stone buildings, predominantly Gothic in style, are built on the Point's high, level plain and in terraced formation along the lower slopes of the surrounding mountains.[55]

WAR DEPARTMENT,

QUARTERMASTER GENERAL'S OFFICE

Specifications for Bed-sheets.

To conform in all respects to the sealed standard sample.

Material.—To be made of fine quality unbleached muslin.

Dimensions.—To be ninety (90) inches long and forty-eight (48) inches wide, when finished.

Workmanship.—To be made with a two (2) inch seam at the top and one (1) inch seam at the bottom, in a neat and substantial manner.

Adopted March 28, 1885.

S. B. HOLABIRD,
Quartermaster General, U. S. A.

1075—F., 1885

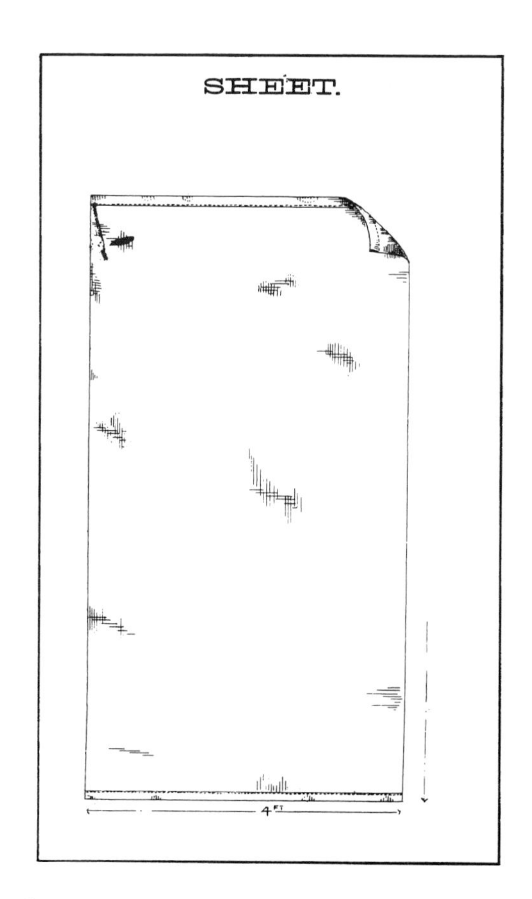

Figure 6.1. Rear View of Old Stone Cadet Barracks, United States Military Academy, West Point, New York, 1885

Enrollment in the cadet corps of the United States Military Academy remained small from the year when the academy first opened, 1802, until the end of the nineteenth century. In 1899, when nineteen-year-old Douglas MacArthur came to West Point as a plebe (freshman), the corps consisted of 332 cadets, less than a tenth of the present generation's yearly enrollments.[56] Until the early 1900s, the gray granite building in Figure 6.1, "grim as a penitentiary,"[57] housed all cadets in the entire corps for the four years of work a plebe needed to become a graduating first classman. In 1965, most of the old building—often called the "Old Stone Barracks" or the "Old First Division Barracks," was torn down, but one four-story section was saved—the one that included the octagonal tower and three bays with central hallways on each floor, each hall with two rooms on either side. Except for the cadet drawing in Figure 6.2, all the scenes of the Military Academy shown on the following pages were made inside this old barracks building.

[Archives of the U.S. Military Academy, West Point, New York]

Figure 6.2. West Point Cadet Room, Sketched by James G. Benton, Class of 1842

The cleverly drawn cartoon in Figure 6.2, showing West Point barracks life as seen through the eyes of a cadet, is the only scene in this section of the book that does not portray some part of the Old Stone Barracks. Cadet James Benton's lovely, informative drawing shows, with youthful zest and exaggeration, a group of cadets supposedly engaged in a great many activities specifically forbidden by the rules of the academy, among them are cooking and eating in a barracks room and smoking, which was not even allowed on the post at that time, much less in the cadet barracks.[58]

The cots on which the two cadets at right are lounging were introduced at the academy in 1838. Before that time, cadets slept on pallets spread on the floors of their rooms. The academy had a severely limited budget and a hard go of it for several years after it opened.

[Archives of the U.S. Military Academy, West Point, New York]

Figure 6.1

Figure 6.2

Figure 6.3

Figure 6.4

Figure 6.3. West Point Cadet Barracks: Room of Cadets Josiah H.V. Field and William R. King, 1860

The name cards "Field" and "King" above the drapery valances at rear indicate that Cadets Field and King shared the West Point barracks room seen in Figure 6.3. Academy records pinpoint the year as 1860. One of the two cadets made this sketch of the room, probably as a classroom assignment, since drawing and drafting were important subjects at the academy throughout the nineteenth century and well into the twentieth. Cadets were taught the principles of drawing and drafting in engineering classes in preparation for planning and drawing blueprints for redoubt and fortifications, battery and artillery sites, and bridges.

The drawing in Figure 6.3 shows clearly the 1838 iron bedsteads still in use at West Point in the 1860s. A gas jet visible on the wall at left, above the study table, provided lighting. Gas lighting was introduced for cadet rooms at West Point in 1858. The big wooden bucket on the low shelf of the washstand held fresh, clean water. Waste water from washing and shaving went into the waste-water bucket on the floor nearby. The water buckets were a part of the room's standard equipment (see Figure 6.5 and 6.6) until the adoption of indoor plumbing, some years later.

The fireplace at right heated the room. The tasseled cords dangling on the chimney breast above the mantel controlled the ventilators.

[Archives of the U.S. Military Academy, West Point, New York]

Figure 6.4 and 6.5. West Point Cadet Barracks: Room of Cadets Frederick A. Hinman and John Pitman, June 1865. Drawn by John Pitman, Class of 1867

Figures 6.4 and 6.5 are the first overall views we have of an entire room in West Point's cadet barracks in the 1800s. The sketches that make that possible were drawn by Cadet John Pitman, class of 1867, who shared this room with Frederick A. Hinman. Pitman was probably making practical application of something learned in his academy drawing classes. Working in the center of the room, he gives us, first, a view of the wall that faces outside, showing the window and the room entrance. Then, still at the center of the room, he turns and draws the opposite, or inside wall, showing the two beds in the sleeping alcove, the washstand, the study table, and the fireplace. Cadets hung their clothing on the walls of the sleeping alcoves (Figure 6.5), where it remained out of sight behind the draperies screening the alcove. The linen presses (Figure 6.4, right rear) contained extra bedding and shirts, with the top shelf reserved for personal items.

[Archives of the U.S. Military Academy, West Point, New York]

Figure 6.5

Figure 6.6. West Point Cadet Barracks: Room of Cadets Isaiah H. McDonald and John B. Kerr, 1870

Figure 6.6 is the first photograph known to have been made of the interior of a West Point cadet room. Cadets McDonald and Kerr (see name sheets below the valances of the draperies of the sleeping alcove) shared these quarters, and Cadet Kerr posed for the photograph, seated on his side of the room. Kerr was Room Orderly at the time, as shown by his name on the tiny tag at the top of the room's center pole—a procedure still observed at West Point today. The valance above the draperies is stamped brass. Mattresses on all cadet beds were to be cleared and folded each day, a practice continued until the 1950s. Two study tables, instead of the usual single one most often seen, appear in this room, and the fabrics used for the draperies and the table covers are of unusual richness. Toweling covers the washstand, and we have a close-up view of the paired wooden water buckets for fresh and waste water.

[Archives of the U.S. Military Academy, West Point, New York]

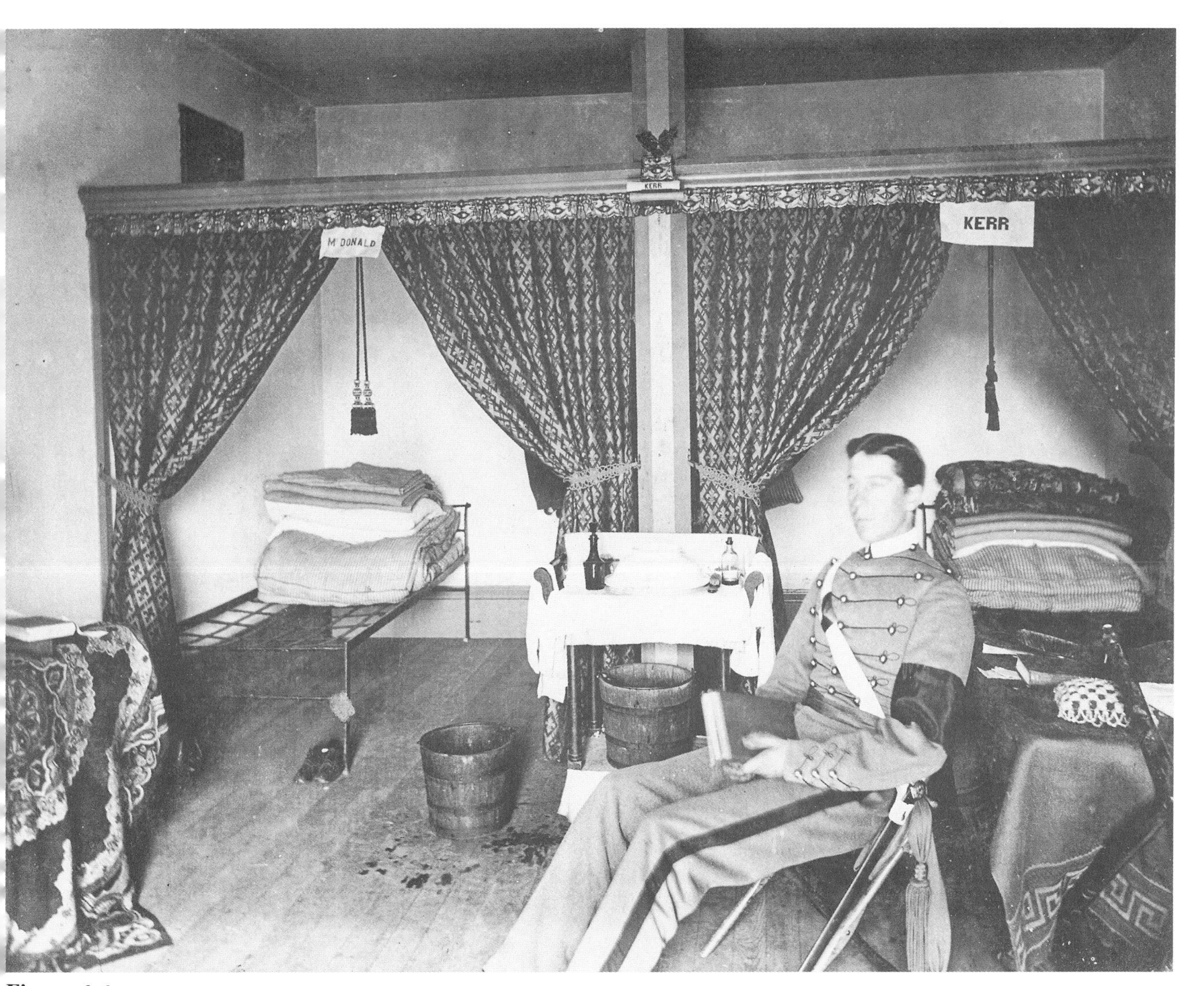

Figure 6.6

Figure 6.7. West Point Cadet Barracks: Tower Suite, 1878. Cadets Frank E. Hobbs and Edward R. Ives

Figure 6.7 shows the "Tower Suite" of the Old First Division Barracks (see Figure 6.1). Beginning in 1851 and continuing past the turn of the century, the Tower Suite was traditionally reserved each year for the top-ranking senior cadet officer of that year's graduating class, the cadet corps' First Captain. Among the corps' most notable First Captains were General Robert E. Lee, class of 1829; General John J. Pershing, class of 1886; and General Douglas MacArthur, class of 1903.[59] Although most of the old barracks that housed the Tower Suite was torn down in 1965, the tower itself, octagonal in shape, can still be seen at West Point today.[60] The four-story end section that included it was kept in tact, with a truncated section of rooms and hallways.

In this 1878 photograph of the Tower Suite, the window treatment is probably blue bunting, a fabric and color the army customarily used for that purpose, matched to blue wool coverings used for desks and tables. A new kind of bunk can be seen through the doorway at right, an iron cot that replaces the type of army bunks in the previous photographs.

Seen dangling from the ceiling's gaslight fixture (center) are "hop" (dance) cards and various other paper notes and mementos, many of them attached to the rubber hose that supplied gas to the student lamp on the table.

[Archives of the U.S. Military Academy, West Point, New York]

Figure 6.8. West Point Cadet Barracks: Unidentified Cadet Room, 1883-1884

This cadet room of the 1880s seems considerably less spacious and gracious than the roomier, more richly furnished one in Figure 6.6. Clearly, the room in Figure 6.8 must be the quarters of plebes who have not yet attained much "standing." Its sparse furnishings give us a clear view of the new iron cots the army started using at West Point in the 1870s, and we notice that the two washstand water buckets here are metal, instead of wood.

[Archives of the U.S. Military Academy, West Point, New York]

Figure 6.7

Figure 6.8

Figures 6.9, 6.10, and 6.11. West Point Cadet Barracks: Room of Cadets Hoover and Lewis, 1916

The three views of the cadet rooms presented in Figures 6.9, 6.10. and 6.11 bring us into the twentieth century at West Point. I include them to help round out the pictorial view presented here of cadet life from the early 1800s through the turn of the century. The room shown in these three photographs is probably closer to what most views regard as "typically West Point." By 1916, cadet rooms were equipped with electric lights (top right, Figures 6.19 and 6.10) and the Model 1893 Quartermaster Bunk. The Quartermaster Bunk was also in use by the this time in army hospitals, but was not found in many photographs of early twentieth-century regular army barracks. As noted with the cadets' use of linen presses in Figures 6.4 and 6.5, the top shelf of the 1916 wall locker in Figures 6.10 and 6.11 might be used for personal items, but academy regulations dictated what was to be stored in the rest of the locker space. Uniform blouses and shirts were to be hung in the full-length space at right. Items of clothing that could be folded were stored, along with extra towels and extra bedding on the adjoining shelves. The white boxes on top of the locker were for stationary. Each cadet's rifle, stacked upright in the wall rack to the right of the lockers, is identified by the name tag of the cadet to whom it has been issued and is protected from dust by being stored in a long black bag. The cadets' "tarbuckets" or shakos—full dress hats—were stored just above the rifles.

[Archives of the U.S. Military Academy, West Point, New York]

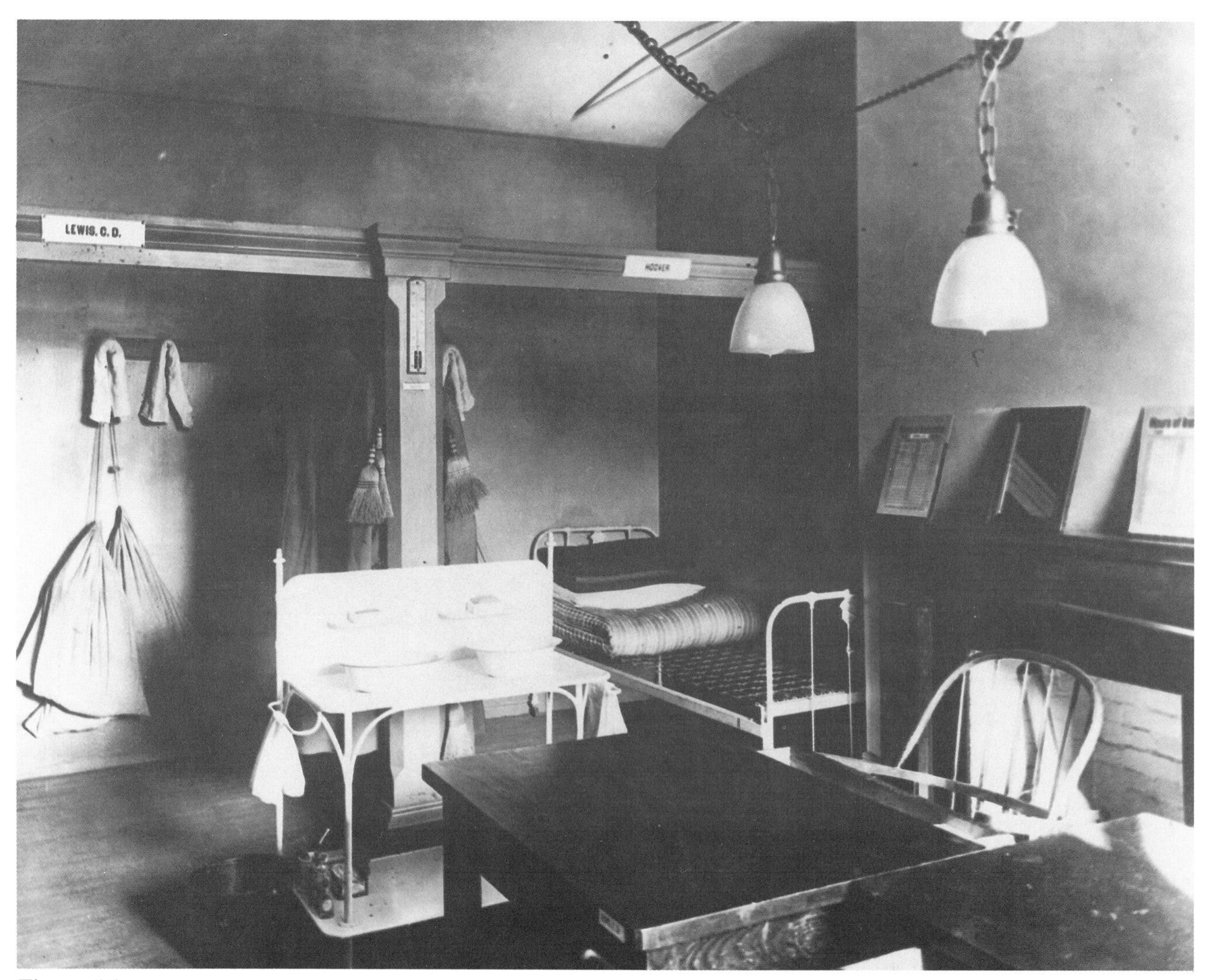

Figure 6.9

Figure 6.10

Figure 6.11

TEMPORARY QUARTERS: CAMP AND FIELD

The human desire to make oneself comfortable, even in the most difficult situations, is one of the things evident in the photographs that follow. Nineteenth-century soldiers in the field—like soldiers of all times and places—devised ingenious ways to carry with them some remarkable items just for that purpose. No sensible soldier would ever take pains to set up all the comforts of home just for the usual campaign-march, one-night stop, of course. But as soon as one reached a destination where one was to stay for any length of time, one made a thorough job of getting as comfortable as possible. The next group of photographs show U.S. Army regulars of the 1800s on field duty, making do—very nicely, thank you—with temporary summer and winter quarters, transient billets, and—always—with tents as "hearth and home."

Tents, their components, their construction, and all the ways one could think of making them more habitable, were important matters to nineteenth-century military men. One of them, Major Henry H. Sibley, of the 2nd U.S. Dragoons, did something about it. In 1856, Major Sibley perfected and patented plans and drawings for a new, improved type of round canvas tent and a small lightweight, portable stove designed to work perfectly in these tents. The stove was easy to transport, safe to use in side inflammable canvas interiors, and thrifty with wood (see Figures 7.3 and 7.6). Both tent and stove were named for their inventor, and both earned popularity with the soldiers because they could be put up and taken down quickly, and they made life in the field on a cold winter day a lot more comfortable.

Sibley tents and Sibley stoves were major items of equipment taken along by troops on the Pine Ridge Campaign in South Dakota during the bitterly cold winter of 1890-1891 (see Figures 7.13 through 7.16). The Pine Ridge Campaign marked the end of the "Indian Wars," decades of sporadic combat between the Plains Indians and the U.S. Army during the last half of the nineteenth century. It was called the "Pine Ridge Campaign" because it took place on and near the Pine Ridge Indian (Sioux) Reservation in western South Dakota. It was a brief campaign, activated in mid-November of 1890, and officially ended on January 15, 1891, when the last of the Sioux chiefs and warriors formally surrendered after a final confrontation with the U.S. Army on the Banks of Wounded Knee Creek on December 29, 1890.[61]

Figure 7.1. Soldiers' Tent, 1715-1720: Bird's Eye View. Engraving by Jean-Antoine Watteau (1684-1721)

The Watteau engraving in Figure 7.1 views, from above, a group of eighteenth-century French infantrymen packed like sausages into a typical enlisted men's tent of the early 1700s. On the march, U.S. Army enlisted men were given straw to sleep on and one blanket apiece for cover. Watteau's French soldiers, lacking even a single blanket, are using their coats for bed-covers. The man nearest the front of the tent (sharp upper left) is fully dressed, with hat, coat, leggings, and sword all on, because he is the group's guard-and-fire watchman and must be prepared to take quick action and rouse the others in the event of any mishaps as they sleep.

[Canadian Park Service, Ottawa, Ontario]

Figure 7.1

Figure 7.2. Tents: Exterior Views, 1765-1774

The engraving in Figure 7.2 shows enlisted men's tents in use by the British Army at about the time of the American Revolution. An A-frame tent with rectangular sides and a back and front shaped like isosceles triangles is shown in the three drawings in the top row of Figure 7.2. This A-frame tent was issued to British infantrymen, and tents just like it were used by U.S. Army enlisted men from Revolutionary War days until the 1840s. In some instances, they were used by the U.S. Army during the Civil War. The first drawing in the top row of Figure 7.2 shows the flat canvas of the A-frame tent stretched on the ground. The center top row drawing shows the A-frame tent's two support poles ("A" and "A") at each end of the tent frame, with the ridgepole ("B") atop them as the tent is set up. The top row drawing at right shows the A-frame tent set up and secured.

The second row of drawings in Figure 7.2 shows a British tent issued to cavalrymen. Like the illustrations in the top row, this first sketch shows the cavalry tent as a piece of canvas spread flat on the ground. The curved part of this tent, seen at the top of the flattened canvas, was intended as a shelter for saddles, bridles, and other riding equipment. Interestingly enough, our Figure 7.1 shows this kind of cavalryman's tent used by French soldiers to shelter *men,* with no space left over for any kind of equipment. The center drawing and the one of the right in the bottom row of drawings show what was called a "bell of arms," a comparatively small tent used to shelter stacked muskets. In the center drawing, we see the canvas flat, and at right, the "bell of arms" set up and secured. American armed forces apparently did not use the "bell of arms."

[From Lochee, "Castrametation in the British Army," London, 1774.

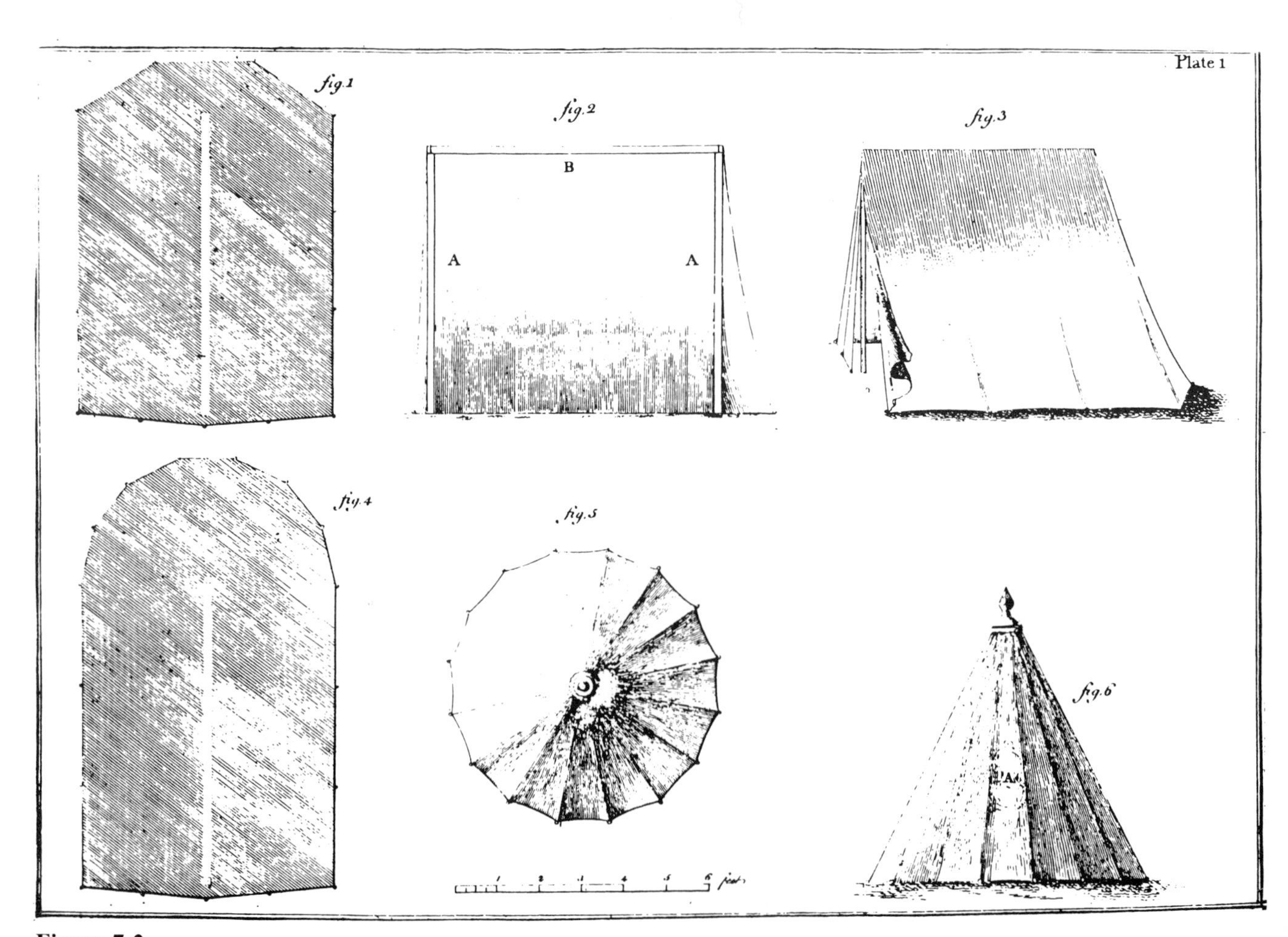

Figure 7.2

Figure 7.3. Interior, Sibley Tent of M.E. Valentine, Company F, 23rd Massachusetts Volunteer Infantry, Annapolis, Maryland, ca. 1862

The early 1860s sketch in Figure 7.3 shows an improved kind of round tent, the Sibley tent, named for its inventor, Major Henry H. Sibley of the 2nd U.S. Dragoons. For use in the new tent, Major Sibley also invented the light, portable, wood-burning stove shown at right in the sketch. Sibley obtained a patent for both tent and stove on April 22, 1856. Both Sibley inventions were instantly popular with the troops. The stove shown here in Figure 7.3 has its short vent pipe thrust through a reinforced exit-hole in the tent wall, with its base set in a small box filled with what appears to be either dirt or sand. Usually, these Sibley stoves were placed under the tent's center support tripod, where they fit neatly and took up a minimum of the tent space. Instead of putting the stove under the tent tripod, however, M.E. Valentine of the Massachusetts Volunteer Infantry has made his tent tripod the center shaft of a handy musket-and-equipment rack, with circular pieces of wood serving as shelves. [Negative Number 165-HV-1, National Archives, Washington, D.C.]

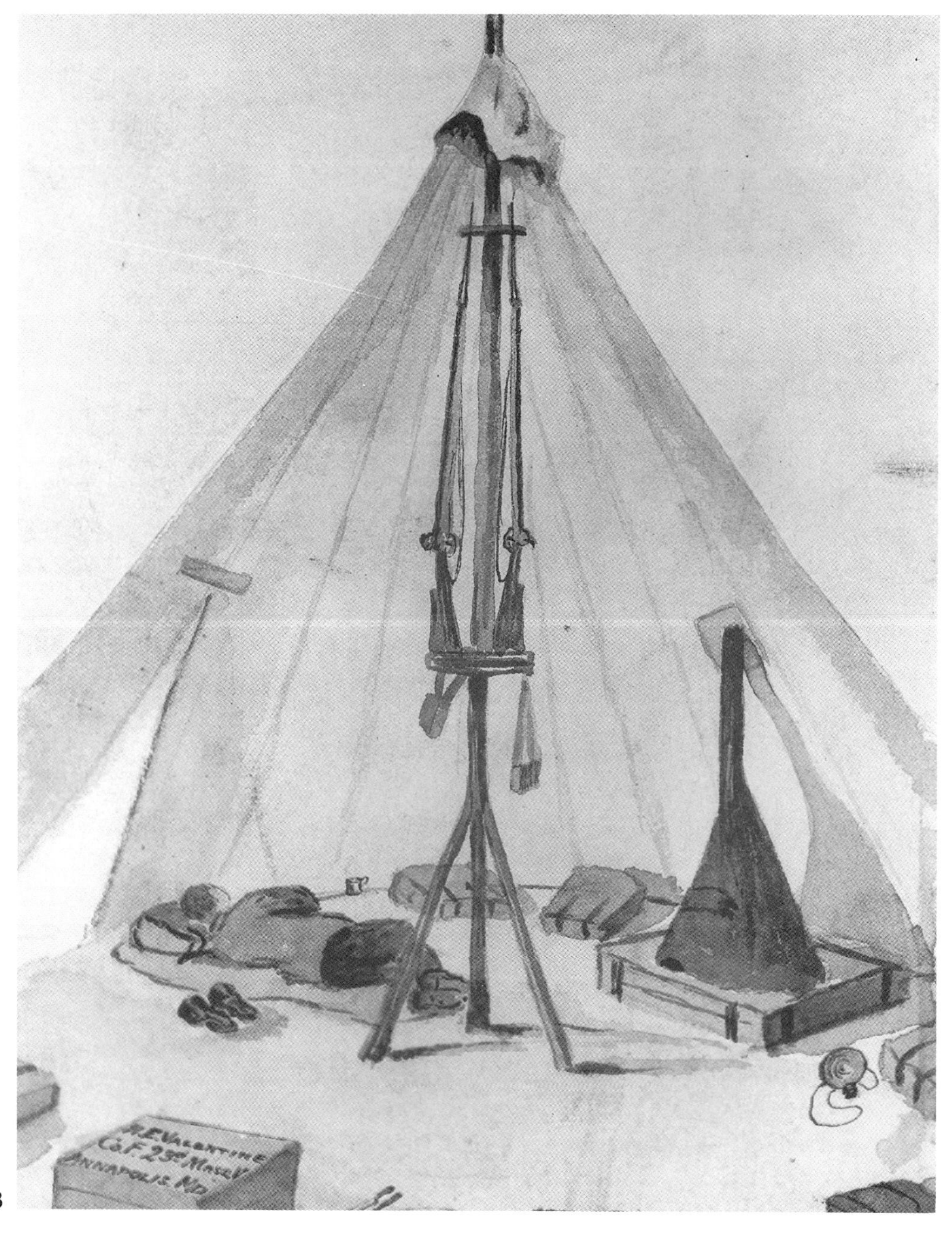

Figure 7.3

Figure 7.4. U.S. Army Officer in Camp, 1862-1865

The photograph in Figure 7.4 shows a third kind of tent used by the military—the wall tent. The wall tent's sides ended about four feet from the ground, and an attached strip of canvas about four feet wide ran perpendicularly from there down to the ground or to the tent floor, it the tent had a floor, and many did. Records indicate that this one did, although it is difficult to see in this photograph. The primary reason for including this picture, however, is to show the view the variety of furniture some army men managed to take along to camp. The tent-dwelling officer, shown in a languorous pose with a letter in one hand, has provided himself with such comforts as the folding cot he is lounging on, a folding table against the far wall of the tent, the typical middle-period "mule-ear" chair on which rests one foot, and a similar chair seen in the right foreground. For the cameraman, this tent-owner has thrown open the tent flap at the front entrance, so that we see his boots, saddle, and other riding equipment stowed on a make-do sawhorse.

[U.S. Signal Corps Photograph, Brady Collection, Negative Number 111-13-6312, National Archives, Washington, D.C.]

Figure 7.4

Figure 7.5. Interior Officer's Tent: Antietam Battlefield, September 1862, CPB, Company K, 19th Regiment, Massachusetts Volunteers

This drawing of the interior of an officer's wall tent actually shows two wall tents, joined together at the ends. The jointure shows clearly at the point where the two sets of floorboards meet. Much to President Lincoln's dismay, the Army of the Potomac remained encamped in the Sharpsburg area for two months after the battle of Antietam was fought on September 17, 1862, and both officers and men made themselves as comfortable as ingenuity and selective scrounging of materials permitted. This officer and his double-tent partner have made themselves quite comfortable. The table at the far end of the tent folds to form a suitcase-like box. Extra tent poles have been put in to hold equipment and blankets. A sword, a holster for a handgun, and a bugle hang among other equipment from the horizontal tent pole. A small candle holder is affixed to the upright pole that rises through the floorboards at the officer's feet, and the classic folding camp chair and stool appear as restful and relaxing as they remain today.

[Antietam National Park, Sharpsburg, Maryland]

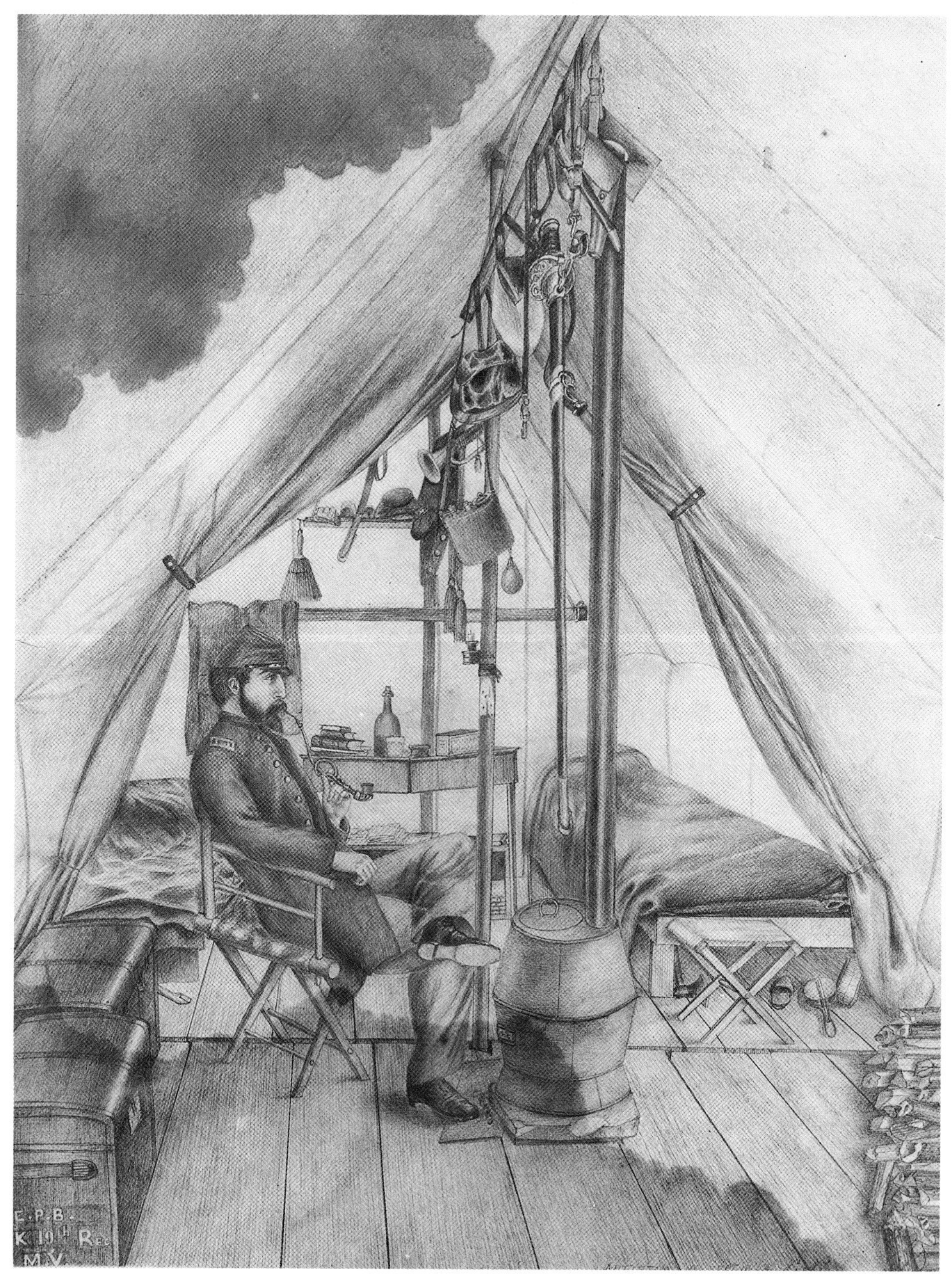

Figure 7.5

Figure 7.6. "My Studio, Rappahannock Station, March 19, 1864, Edwin Forbes"

Mr. Forbes' title for this drawing of his wall tent appears in the lower right corner, with the location and date dimly visible at lower left. Though not an officer, Forbes seems to have traveled and lived in much the same style as army officers of the 1860s. Forbes was an artist, on assignment as war correspondent for *Harper's Weekly*. Winslow Homer also covered the Civil War as correspondent and artist for *Harper's Weekly,* and Frederick Remington did similar work for the Hearst papers during the Spanish-American War, at century's end.

The Forbes drawing in Figure 7.6 shows a kind of folding camp stool different from that in Figure 7.5. The one in Figure 7.6 has three legs instead of four. Use of the bayonet as a candle holder, as seen here between the table and the Sibley stove's sandbox base was common among the soldiers. The blade of the bayonet was stuck into the ground, and the candle was stuck into the metal fitting that held the bayonet on the rifle barrel for its military purpose in combat. This tent has no floor, but straw has been strewn about on the ground, much like the rushes in medieval times. The Sibley stove rests safely in its box, out of the way of people entering and leaving the tent.

[Negative Number 12221-1841, Library of Congress, Washington, D.C.]

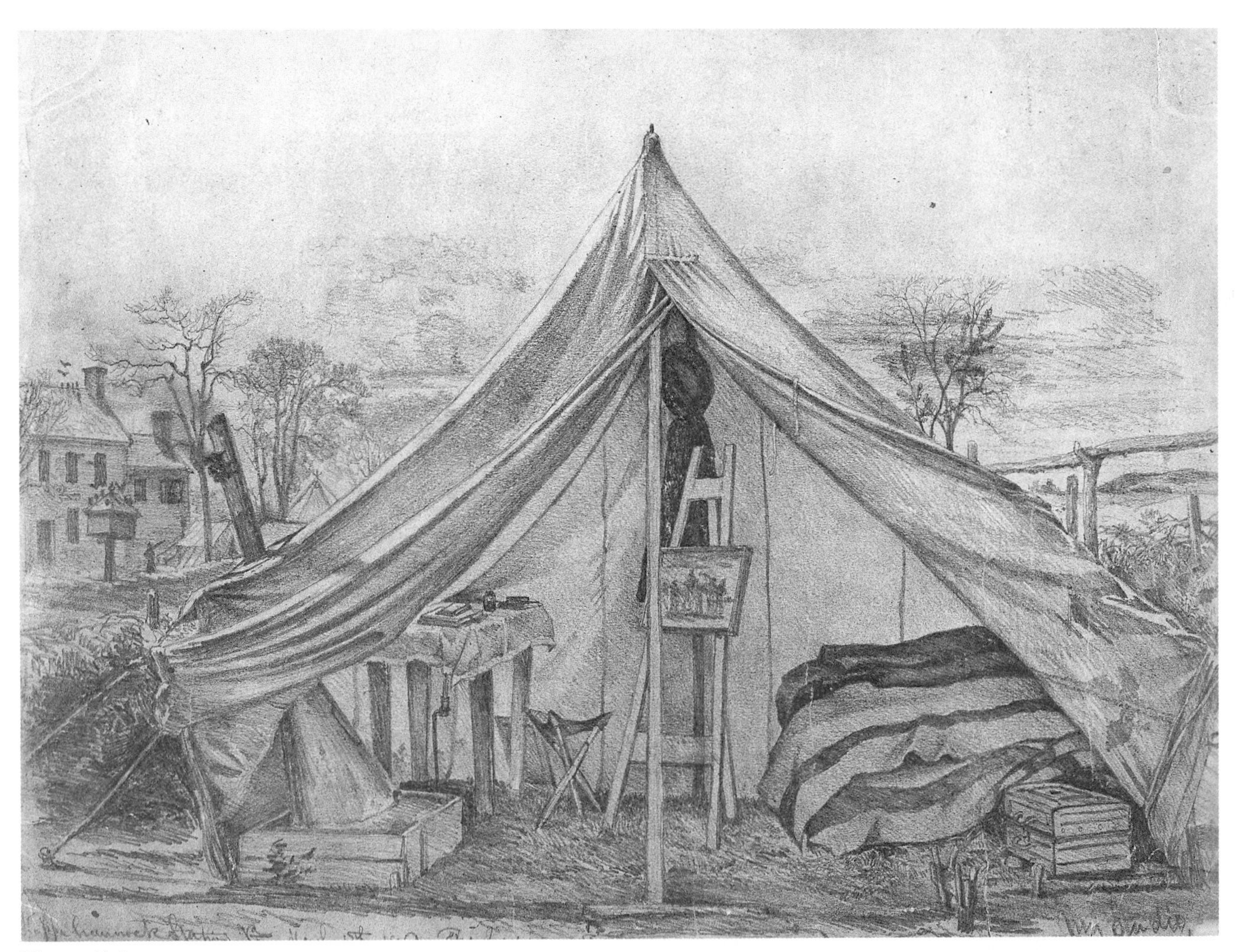

Figure 7.6

Figure 7.7. Tent Interior: Side B of a Stereoscopic Pair

Like Figure 7.5, the Figure 7.7 view of a tent interior of the 1860s shows two adjoining wall tents moved together to make one large space. Here, army blankets are used as carpets, a usage that became quite common in more permanent army posts after the Civil War ended. The officers sharing this expanded billet made it into one of the most comfortable-looking camp quarters shown. The two well-built folding chairs at left are quite handsome. The table against the rear wall, the washstand with its big water bucket, and the bedstead are good wooden pieces, sturdily made. The canvas folding chair at lower right looks as timelessly inviting as it would be on a back patio today.

[Negative Number 111-B-3650, National Archives, Washington, D.C.]

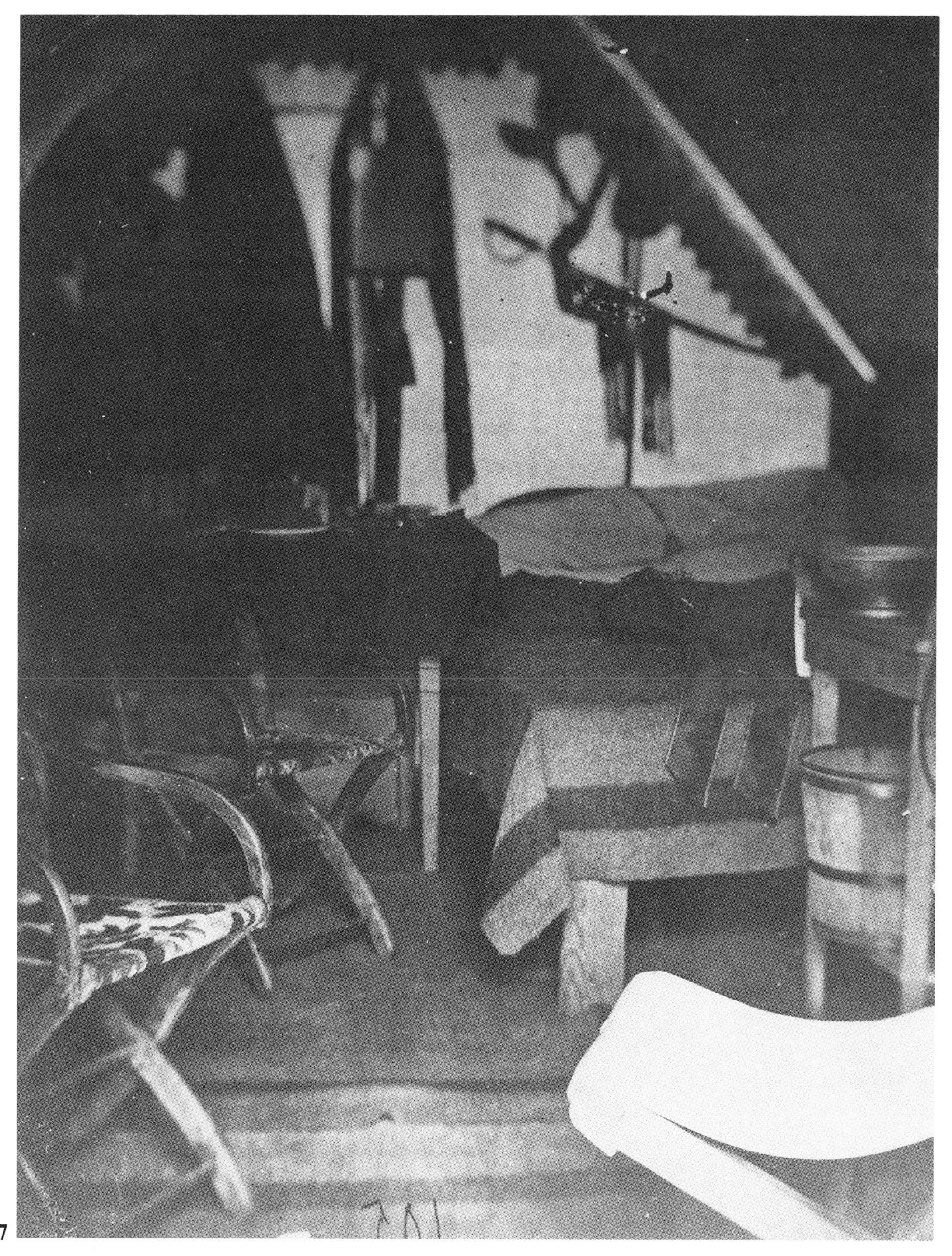

Figure 7.7

Figure 7.8. Winter Quarters of Battery E, 1st Rhode Island Light Artillery, 1861-1865

The room shown in Figure 7.8 is part of the winter quarters of the 1st Rhode Island Light Artillery's Battery E. The battery's commanding officer and the senior non-commissioned officer (neither of whom is identified) have apparently just finished a meal and are posing for the cameraman with several enlisted men. Behind the three men at right, part of a large fireplace topped by a mantel can be seen. Crossed swords and old pistols decorate the wall above the mantel. The large sheets of old paper on the rear wall carry printing of some sort, but it is difficult to tell whether these papers are maps or merely old newspaper pages pasted on the wall to cover cracks and cut down drafts. Judging by its size and furnishings, this room may have been the office of Battery E, as well as the officer's dining room.

[Massachusetts Commandery, Military Order of the Loyal Legion, Volume 113, p. L5813, U.S. Army Military History Institute, Carlisle, Pennsylvania]

Figure 7.9. "This Cruel War! Roughing It at Arlington House, Virginia." Sketch by Charley Morgan, ca. 1862

The blithe-spirited sketch in Figure 7.9 shows and upstairs bedroom at Arlington House, also called the Custis-Lee Mansion, near Alexandria, Virginia, in the early 1860s. Arlington House was, for many years, the home of Confederate General Robert E. Lee and Mrs. Lee. Mrs. Lee, a great-granddaughter of Martha Custis Washington, inherited the mansion. Early in the Civil War, the Lees moved to Richmond, Virginia, capital of the Confederacy, and abandoned the house, which was later occupied and used as headquarters by the Union Army. Much of the Lee's furniture remained at the mansion and provided a very atypical wartime environment for soldiers like the young man so obviously enjoying himself there in 1862. The drawing must have been made in the summer—the windows appear to be open, the shrubbery in full leaf, what appear to be a bottle and a bowl are the only things visible in the ash-free fireplace, and there are no heavy winter bed-hangings in evidence. Some two hundred acres of the Arlington House estate were set aside in 1864 to be used as a national cemetery. Today, the mansion and immediate grounds are maintained as a memorial to General Lee.

[Custis-Lee Mansion, National Park Service]

Figure 7.8

Figure 7.9

Figure 7.10

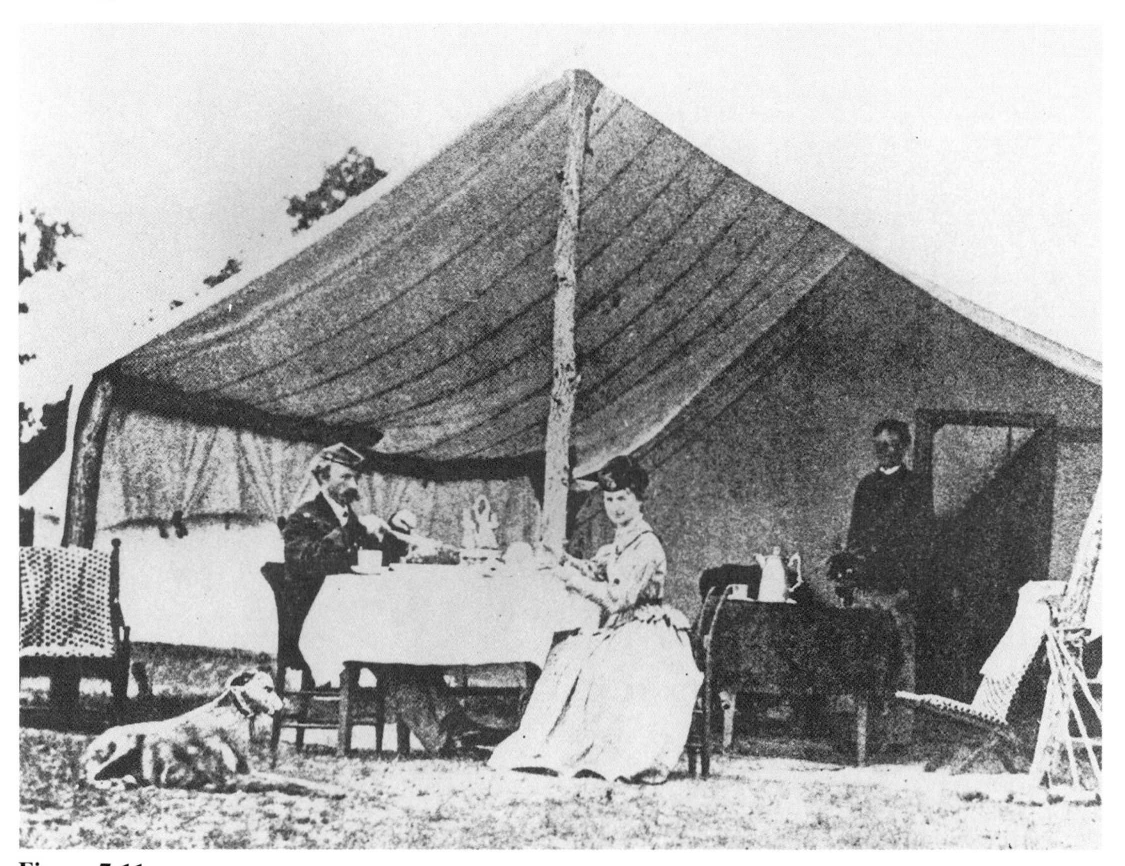

Figure 7.11

Figure 7.10. Surgeon McKay's Quarters, Virginia, 1864

The room seen in Figure 7.10 was part of an abandoned house used as field quarters by a group of U.S. Army officers in Virginia in 1864. Records indicate that a Surgeon McKay was more or less in charge of the group and that the five men pictured were probably doctors. Although they appear a solemn group before the camera, they have gone to some lengths to bring a touch of whimsical cheer to this room, which probably served as a combination field office and recreation room. A field desk that is clearly in use is seen at right rear, on two small tables. A checkerboard is propped up in the center of the mantelpiece. Several musical instruments are prominently displayed, side by side with the weapons of war. A card game that looks very much as if it had been carefully staged for the camera is supposedly in progress, and a great display of what was surely labeled "Medicinal Alcohol" appears on the card table.

The makeshift furniture was probably gathered from other rooms in the house or from the neighborhood nearby. The graffiti above the mantel is the doctor's own, and the card table appears to be covered by a rubber blanket.

[Negative Number LC-B8184-1024, Library of Congress, Washington, D.C.]

Figure 7.11. Lieutenant Colonel George Armstrong Custer and Mrs. Custer in Camp, 1867

In Figure 7.11, Lieutenant Colonel George Armstrong Custer and Mrs. Custer dine alfresco, under a tent fly, a sort of porch attached to their sleeping tent, which can be seen at rear. This photograph was probably made at the Custer's summer quarters at Fort Wallace, Kansas, where Colonel Custer was stationed in the late 1860s. The Custers' striker—an enlisted man assigned routinely to officers' families as butler/houseman—stands in the background, at right rear, with his eye on the teapot. The family pet rests in the foreground. Sturdy wooden tables and chairs furnish this outdoor dining area in camp, and white table linens grace the table for the colonel's dining.

[Kansas State Historical Society, Topeka, Kansas]

Figure 7.12. Quarters of Second Lieutenant Granger Adams, 5th U.S. Artillery, Camp Summerville, South Carolina, 1877

Figure 7.12 reveals that sometimes lieutenants in the nineteenth-century army enjoyed almost as many creature comforts as did officers of much higher rank. Lieutenant Granger Adams (the big man seated at far right) has acquired a well-made wooden floor for his tent, which has fine cross-ventilation to catch the summer breeze. And we can see, hanging from a frame above the cot (behind Lieutenant Adams and the man next to him) the good, sturdy mosquito netting he was fortunate also in acquiring—mosquitos show no respect for a uniform, whatever the rank of its wearer. As in the Custers' summer quarters (Figure 7.11), there is, in this officer's tent, a surprising lot of good, solid, wooden furniture—a very nice table at right, with a handsome kerosene lamp on it, three good straight chairs, both the folding kind, a big chest of drawers at left, and a wooden washstand furnished with a graceful basin and a big wooden water bucket.

[Negative Number 1005, B.F. Browne Collection, National Archives, Washington, D.C.]

Figure 7.12

Figures 7.13, 7.14, 7.15, and 7.16. Sibley Tents of Officers and Men on the Pine Ridge Campaign in South Dakota, 1890-1891

The Pine Ridge Campaign, so named because it took place on and around the Pine Ridge Indian Reservation in South Dakota, was mobilized in the winter of 1890-1891. The last of many campaigns fought between the Plains Indians and the U.S. Army in the decades after the Civil War, the Pine Ridge Campaign ended January 15, 1891, just after the Battle of Wounded Knee.

Dakota winters are fearfully cold, and the army's Sibley tents, like the ones seen in Figures 7.13 through 7.16, with the quickly set-up Sibley stoves, made ideal winter campaign quarters for the officers and the men of the 1st and the 8th U.S. Cavalry and the Buffalo Soldiers of the 25th U.S. Infantry, all camped along the banks of Wounded Knee Creek, near the Pine Ridge Reservation.

In Figure 7.13, officers of the 1st Cavalry thaw out around the little stove as a dispatch is read, while water heats for tea. In Figure 7.14, two of the 25th Infantry officers enjoy the stove's warmth as they catch up on map-reading while the soup warms.

Figure 7.15 shows the kitchen tent of the 25th Infantry, with its small rectangular cook stove, set up back-to-back with the Sibley stove. Hardtack, the nineteenth-century army's durable, dependable, fast-food staple, along with other nonperishable foodstuffs were carried on the campaign trail in the company mess chest, seen with its lid open between the two seated cooks. Even on campaign, some of the niceties were observed; note the brightly figured oilcloth with covers the kitchen work table.

In Figure 7.16, men of the 25th Infantry show the photographer sleeping arrangements in the Sibley tent—feet towards the fire, head towards the circular tent wall, and keep your earmuffs on. It was a universal old army wheeze that, when all the troops that a tent could hold for sleeping crowded in, if one wanted to turn over, all he had to do was yell "Spoon left!" or "Spoon right!" and everybody in the tent turned over, in the direction called for.

[Negative Numbers 111-SC-104138, 111-SC-83639, 111-SC-83763, and 111-SC-83764, National Archives, Washington, D.C.]

Figure 7.13

Figure 7.14

Figure 7.15

Figure 7.16

Figure 7.17. Campaign Mess Tent, Fort Leavenworth, Kansas, ca. 1890

Figure 7.17 shows a big wall tent with a wooden floor that was a prototype for campaign mess tents of the late 1800s. A tent of this size would not be set up for a small number of men nor for a very brief stay in any one area, so it was usually furnished with the same kind of heavy mess tables and benches used in more permanent army dining structures. A good bit of ironstone tableware is seen in this mess tent, most of it probably bought by the company or with regimental funds.

[Kansas State Historical Society, Topeka, Kansas]

Figure 7.17

APPENDIX ONE

Locations of Forts

A wide variety of United States Army posts are represented by the photographs in this book, and their locations are scattered across the United States. Those wishing a closer identification of the posts with cities or modern locales might find this list of some use, but only those posts depicted in this text are included. Information on the locations is from Francis Paul Prucha, *A Guide to the Military Posts of the United States* and Francis B. Heitman, *Historical Register and Dictionary of the United States Army*.

Angel Island	San Francisco Harbor, CA
Bencia	San Francisco Harbor, CA
Camp Pilot Butte (Camp Rock Springs)	Rock Springs, WY
Columbus Barracks	Columbus, OH
Fort Abraham Lincoln	near Bismark, ND
Fort Adams	Newport Harbor, RI
Fort Alcatraz	San Francisco Harbor, CA
Fort Assiniboine	near Havre, MT
Fort Barrancas	Pensacola Harbor, FL
Fort Bayard	near Silver City, NM
Fort Bliss	El Paso, TX
Fort Bowie	east of Wilcox, AZ
Fort Bridger	west of Rock Springs, WY
Fort Brown	Brownsville, TX
Fort Clark	Brackettsville, TX
Fort Columbus	Governors Island, NY
Fort Craig	west bank of the Rio Grande, NM
Fort Crook	Omaha, NE
Fort Custer	Hardin, MT
Fort D.A. Russell	Cheyenne, WY
Fort Dodge	Dodge City, KS
Fort Douglas	near Salt Lake City, UT
Fort Ethan Allen	near Burlington, VT
Fort Grant	near Wilcox, AZ
Fort Hamilton	Long Island, NY
Fort Huachuca	Huachuca City, AZ
Fort Jefferson	west of Key West, FL
Fort Keogh	Miles City, MT
Fort Larned	near Larned, KS
Fort Lawton	WA
Fort Leavenworth	Leavenworth, KS
Fort Logan	near Denver, CO
Fort McHenry	Baltimore, MD

Fort McIntosh	near Laredo, TX
Fort McPherson	Atlanta, GA
Fort McPherson	near North Platte, NE
Fort Mackinac	Mackinac Island, MI
Fort Marcy	Santa Fe, NM
Fort Mason	San Francisco, CA
Fort Meade	east of Deadwood, SD
Fort Mifflin	near Philadelphia, PA
Fort Monroe	Hampton Roads, VA
Fort Myers	Arlington, VA
Fort Niagara	Youngstown, NY
Fort Porter	Buffalo, NY
Fort Preble	Harbor Portland, ME
Fort Riley	near Junction City, KS
Fort Ringgold	Rio Grande City, TX
Fort Robinson	near Crawford, NE
Fort Sam Houston	San Antonio, TX
Fort Scott	Fort Scott, KS
Fort Shaw	near Great Falls, MT
Fort Sheridan	Highland Park, IL
Fort Sill	Lawton, OK
Fort Snelling	Minneapolis/St. Paul, MN
Fort Stanton	Fort Stanton, NM
Fort Townsend	Fort Townsend, WA
Fort Trumbull	near London, CT
Fort Union	near Waltrons, NM
Fort Verde	Camp Verde, AZ
Fort Walla Walla	Walla Walla, WA
Fort Warren	Boston Harbor, MA
Fort Wayne	near Detroit, MI
Fort Wingate	east of Gallup, NM
Fort Wood	New York Harbor, NY
Fort Yellowstone	Mammoth Hot Springs, WY
Jackson Barracks	New Orleans, LA
Jefferson Barracks	near St. Louis, MO
Key West	Key West, FL
Madison Barracks	Sachets Harbor, NY
Mount Vernon Barracks	Mount Vernon, AL
Plattsburg Barracks	Plattsburg, NY
Rosebud Indian Agency	Rosebud, SD
San Carlos	San Carlos, AZ
Sidney Barracks	Sidney, NE
Vancouver Barracks	near Portland, OR
Washington Barracks	Washington, DC
Whipple Barracks	north of Prescott, AZ
Willets Point	Willets Point, NY

APPENDIX TWO

The following is a section reproduced out of *The Prairie Traveler: A Hand-Book for Overland Expeditions.* Pages 144—150 of this book have been included in order for readers to have a closer and more in-depth look at the types of portable and collapsible furniture that was available to nineteenth-century servicemen. These types of furniture were most useful to those who were frequently moving from place to place, and did not have the time or the funds to construct a permanent residence with expensive furnishings. They were easily transported, and many times served a variety of purposes. Simultaneously, they provided the soldier with the comforts of home, at very little cost.

THE

PRAIRIE TRAVELER.

A HAND-BOOK FOR

OVERLAND EXPEDITIONS.

WITH MAPS, ILLUSTRATIONS, AND ITINERARIES OF THE PRINICIPAL ROUTES BETWEEN THE MISSISSIPPI AND THE PACIFIC.

BY RANDOLPH B. MARCY,

CAPTAIN U. S. ARMY.

enough to admit of the tent's being suspended by ropes attached to the apex. This method dispenses with the necessity of the central upright standard.

When the weather is very cold, the tent may be made warmer by excavating a basement about three feet deep, which also gives a wall to the tent, making it more roomy.

The tent used in the army will shelter comfortably twelve men.

Captain G. Rhodes, of the English army, in his recent work upon tents and tent-life, has given a description of most of the tents used in the different armies in Europe, but, in my judgment, none of them, in point of convenience, comfort, and economy, will compare with the Sibley tent for campaigning in cold weather. One of its most important features, that of admitting of a fire within it and of causing a draught by the disposition of the wings, is not, that I am aware, possessed by any other tent. Moreover, it is exempt from the objections that are urged against some other tents on account of insalubrity from want of top ventilation to carry off the impure air during the night.

CAMP FURNITURE.

The accompanying illustrations present some convenient articles of portable camp furniture.

Camp Chair No. 1 is of oak or other hard wood. Fig. 1 represents it opened for use; in Fig. 2 it is closed for transportation. *A* is a stout canvas,

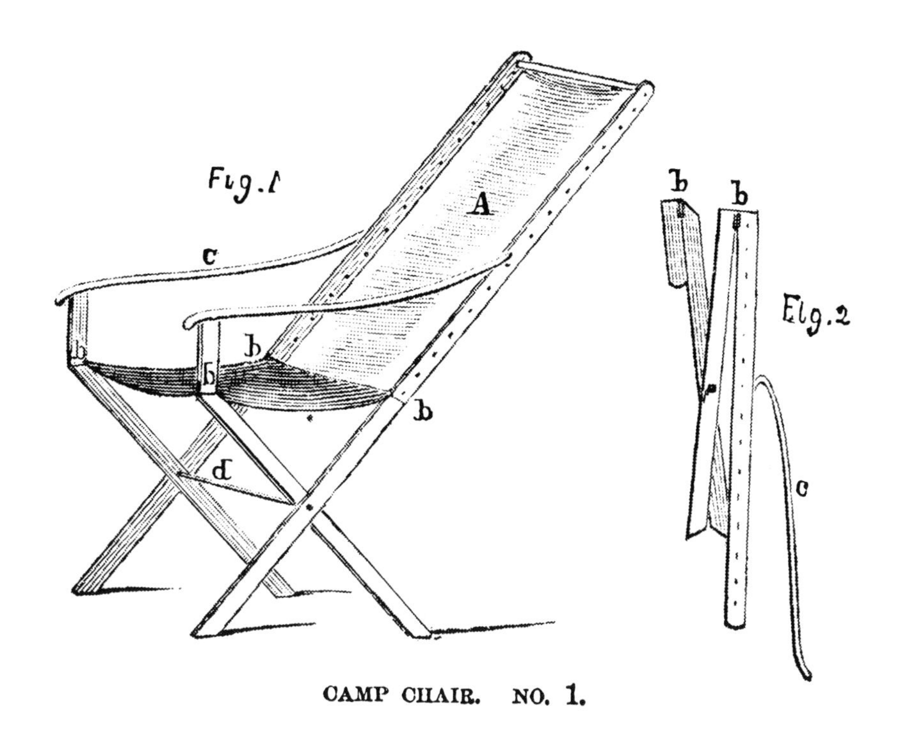

CAMP CHAIR. NO. 1.

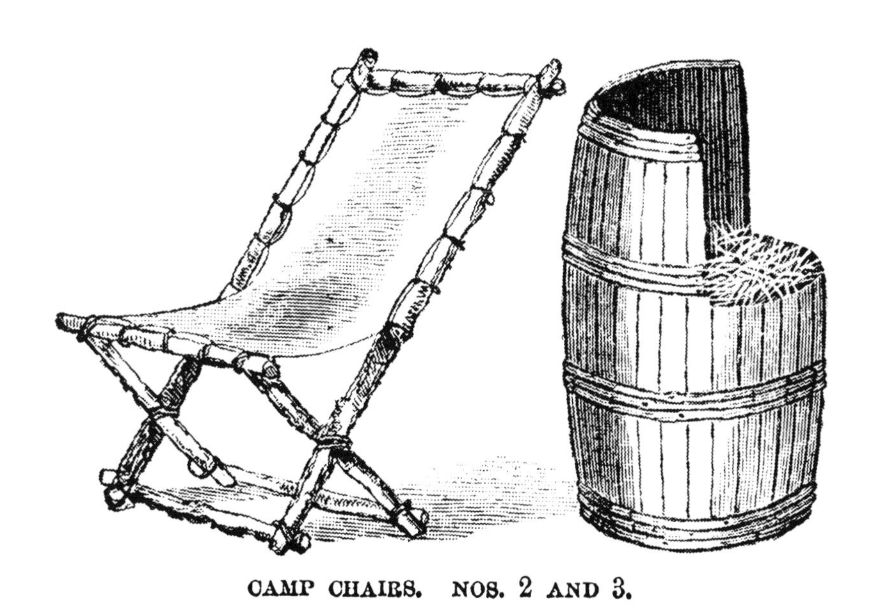
CAMP CHAIRS. NOS. 2 AND 3.

K

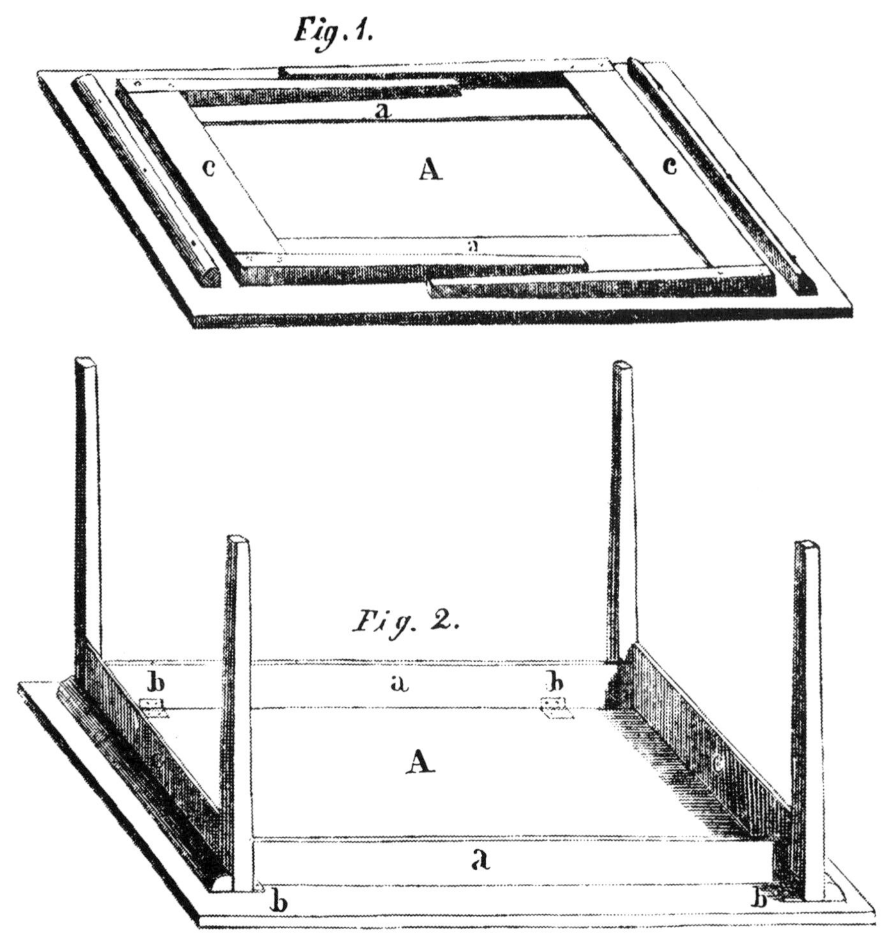

CAMP TABLE

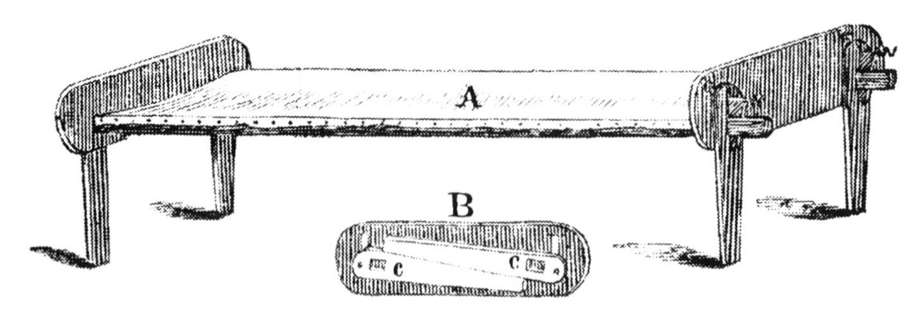

FIELD COT. NO. 1.

forming the back and seat; *b*, *b*, *b* are iron butt-hinges; *c*, *c* are leather straps, one inch and a quarter wide, forming the arms; *d* is an iron rod, with nut and screw at one end.

Camp Chair No. 2 is made of sticks tied together with thongs of buckskin or raw hide.

Camp Chair No. 3 is a very comfortable seat, made of a barrel, the part forming the seat being filled with grass.

Camp Table. Fig. 1 represents the table folded for transportation; in Fig. 2 it is spread out for use. *A* is the top of the table; *a*, *a* are side boards, and *c*, *c* are end boards, turning on butt-hinges, *b*, *b*, *b*.

Field Cots. In No. 1, *A* represents the cot put up for use; *B*, the cot folded for transportation. The legs turn upon iron bolts running through the head and foot boards; they are then placed upon the canvas, and the whole is rolled up around the side pieces. In No. 2 the upper figure represents the cot put up for use; the lower shows it folded for transportation. *A* is a stout canvas; *b*, *b* are iron butt-hinges; *c*, *c*, the legs; *d*, *d*, leather straps, with buckles, which hold the legs firm; *f*, *f*, ends, which fold upon hinges; *g*, *g*, cross-bars from leg to leg. This cot is strong, light, and portable.

Camp Bureau. This cut represents two chests, *A*, *A*, with their handles, *a*, *a*; the covers taken off, they are placed one upon the other, and secured by the clamps *B*, *B*; *d* shows the division between the two chests. When it is to be transported, the

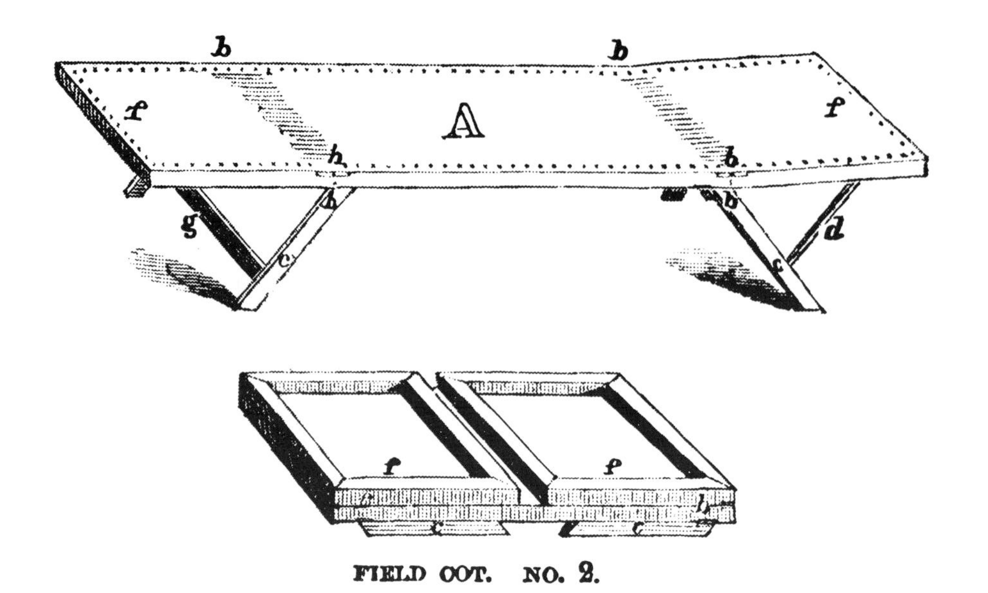

FIELD COT. NO. 2.

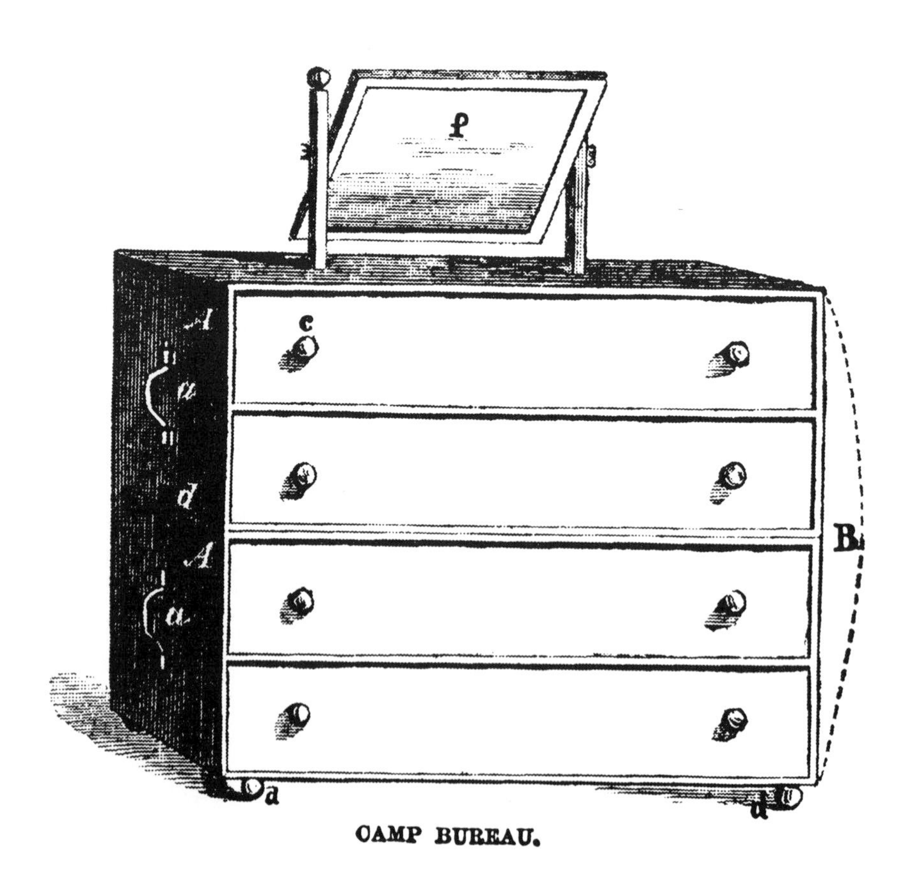

CAMP BUREAU.

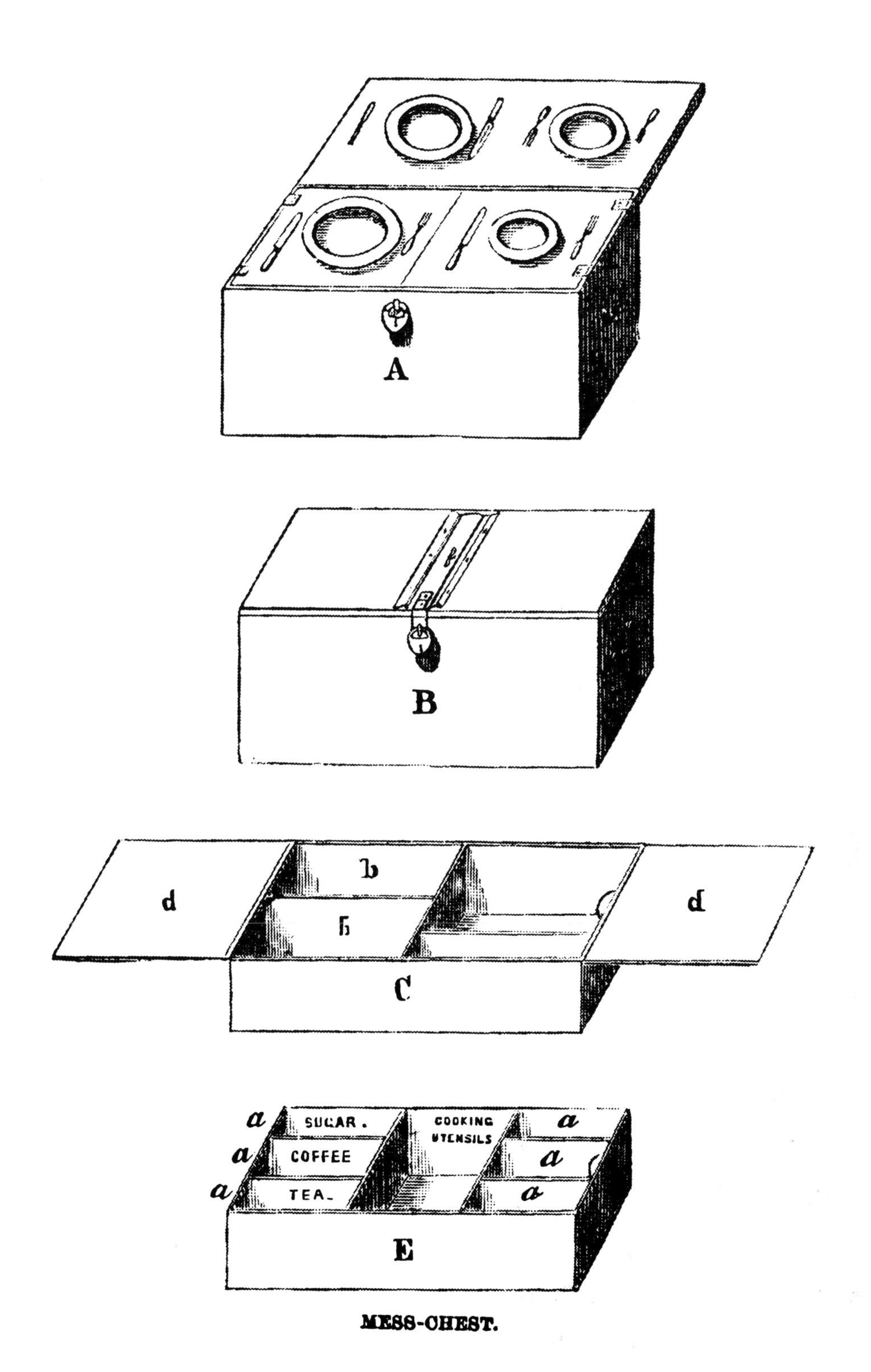

MESS-CHEST.

knobs, *c*, are unscrewed from the drawers, the looking-glass, *f*, is removed, the drawers are filled with clothing, etc., and the lids are screwed on.

Mess-chest. *A* represents the chest open for table; *B* is the same closed; *C* is the upper tray of tin, with compartments, *b*, *b*; *E* is the lower wooden tray, divided into compartments, *a*, *a*, for various purposes, and made fast to the bottom of the chest; *d*, *d* are lids opening with hinges; *f* (in figure B) is a wooden leg, turning upon a hinge, and fitting snugly between two pieces of wood screwed upon the cover.

LITTERS.

Should a party traveling with pack animals, and without ambulances or wagons, have one of its members wounded or taken so sick as to be unable to walk or ride on horseback, a litter may be constructed by taking two poles about twenty feet in length, uniting them by two sticks three feet long lashed across the centre at six feet apart, and stretching a piece of stout canvas, a blanket, or hide between them to form the bed. Two steady horses or mules are then selected, placed between the poles in the front and rear of the litter, and the ends of the poles made fast to the sides of the animals, either by attachment to the stirrups or to the ends of straps secured over their backs.

The patient may then be placed upon the litter, and is ready for the march.

NOTES

[1] Frances M. Roe. *Army Letters from an Officer's Wife* (Lincoln, NE: University of Nebraska Press, 1981), p. 89.

[2] Elizabeth B. Custer. *Boots and Saddles,* 1885 (Norman, OK: University of Oklahoma Press, 1961), p. 102.

[3] Alice Blackwood Baldwin. *An Army Wife on the Frontier: The Memoirs of Alice Blackwood Baldwin, 1876-1877,* edited by Robert C. Carriker and Eleanor R. Carriker (Salt Lake City, UT: University of Utah Library, 1975), pp. 30, 34, 35, 38, 48, 56.

[4] "Letters of Caroline Frey Winne from Sidney Barracks and Fort McPherson, Nebraska, 1874-1878." Edited by Thomas R. Buecker, *Nebraska History Magazine,* Vol. 62, No. 1 (Spring 1982), pp. 7,8.

[5] Etta White to W.F. White, January 4, 1878. Fort Larned Collection, Fort Larned, Kansas.

[6] John G. Bourke. *On the Border with Crook.* (New York: Charles Scribner's Sons, 1891), pp. 6-8.

[7] *Army and Navy Journal,* January 11, 1868, p. 330.

[8] J.C. Furnas. *The Americans: A Social History of the United States, 1587-1914.* (New York: G.P. Putnam's Sons, 1969), pp. 671 and 429; William Seale, *Recreating the Historic House Interior.* (Nashville: American Association for State and Local History, 1979), pp. 71-72; Daniel J. Boorstin, *The Americans: The Democratic Experience.* (New York: Random House, 1973), p. 43.

[9] William Seale. *The Tasteful Interlude: American Interiors Through the Camera's Eye,* 1860-1917, 2nd ed. (Nashville: American Association for State and Local History, 1981), pp. 117, 122.

[10] Custer, p. 102

[11] Catharine E. Beecher and Harriet Beecher Stowe. *American Woman's Home: Or, Principles of Domestic Science, 1869.* (Hartford, CT: Stowe-Day Foundation, 1975, Figure 45, p. 97.

[12] John S. Billings. *Assistant Surgeon General, U.S. Army Report on Hygiene of the United States Army,* Circular No. 8, Surgeon General's Office. (Washington, DC: Government Printing Office, 1875), p. ix.

[13] Autobiography of an English Soldier in the United States Army. (New York: Stringer and Townsend, 1853), p. 34.

[14] John S. Billings. *Assistant Surgeon General, U.S. Army Report on Barracks and Hospitals,* Circular No. 4, Surgeon General's Office (Washington, DC: Government Printing Office), p. xvi.

[15] Billings. *Report on Hygiene of the United States Army,* p. xviii.

[16] Inspection Reports of Colonel George Croghan, 1826, Office of the Inspector General, Record Group 159, Vol. 2, National Archives, Washington, DC, 1826-1845. Information in this caption is drawn from a general reading of Croghan's reports from 1826 to 1845. Another excellent source is Francis Paul Prucha, *Army Life on the Western Frontier* (Norman, OK: University of Oklahoma Press, 1958).

[17] *Columbia Encyclopedia,* 4th ed., s.v. "Lighting," "Edison, Thomas Alva"; Howard L. Hurwitz, *An Encyclopedic Dictionary of American History,* revised and updated (New York: Pocket

Books/Simon and Schuster, 1963; New York: Washington Square Books, 1970, 1974); Boorstin, pp. 534-536.

[18] Samuel Eliot Morison. *Oxford History of the American People* (New York: Oxford University Press, 1965), pp. 226, 236, 653, 674-675; *Columbia Encyclopedia,* 4th ed., s.v. "Negro"; Edward S. Barnard (ed.), *Reader's Digest Association Story of the Great American West* (Pleasantville, NY: Reader's Digest Association, 1977), p. 229.

[19] Jules S. Billard (ed.). *The World of the American Indian* (Washington, DC: National Geographic Society, 1974), p. 342; Dee Brown, *Bury My Heart at Wounded Knee* (New York: Holt, Rinehart and Winston, Inc., 1971; New York: Bantam Books, 1972), pp. 237, 239; *Reader's Digest,* p. 229.

[20] *Reader's Digest,* p. 229.

[21] *Colombia Encyclopedia,* s.v. "Negro," "Integration"; Morison, *Oxford History of the American People,* p. 985

[22] Morison, pp. 750-751; Brown p. 106, 294, 310, 409, 411; *Reader's Digest,* pp. 221, 228, 231, 234, 241.

[23] *Reader's Digest,* pp. 221, 228, 231, 234, 241; *The World of the American Indian,* pp. 332, 338, 366; Morison, p. 751; Brown, pp. 108-109, 168, 275-276, 317, 379-380, 383-385.

[24] *The World of the American Indian,* pp. 278, 281, 318-319, 329, 330, 336, 339; Morison, pp. 225-226; *Reader's Digest,* pp. 8, 224, 242, 243, 268-269.

[25] *Reader' Digest,* pp. 238, 240, 243-244; Brown, pp. 106, 294-295, 310, 372, 409, 411.

[26] Dave Richard Palmer and James W. Stryker. *Early American Wars and Military Institutions* (West Point Military History Series; Avery Publishing Group, Inc., 1984), pp. 68-69, 75; *Reader's Digest,* pp. 209, 221, 228, 231, 234, 241; Brown, pp. 106, 108-109, 168, 266, 294, 310, 379-380, 383-385; *World of the American Indian,* pp. 338, 366.

[27] Brown, pp. 108-109, 168, 286-287, 291; *Colombia Encyclopedia,* s.v. "Pawnees", p. 516.

[28] *Reader's Digest,* p. 221; Brown, p. 349.

[29] *Reader's Digest,* pp. 221, 228, 241; Brown, pp. 275-276, 379-380, 383-385; *Colombia Encyclopedia,* s.v. "Crook, George."

[30] *World of the American Indian,* p. 338; *Reader's Digest,* p. 234

[31] *World of the American Indian,* p. 368, 359.

[32] *Webster's Third New International Dictionary of the English Language,* Unabridged (Springfield, MA: G. & C. Merriam Company, 1971), s.v. "Boater," "kady, katy, cady," "sailor hat"; Eric Partridge, *A Dictionary of Slang an Unconventional English,* 7th edition (New York: Macmillan Company, 1970), s.v. "cady," "kadi," notes that "kadi" or "cady" was an English work for "hat in the 1880s and for a straw hat in particular in the early 1900s.

[33] Furnas, p. 896.

[34] *Story of the Great American West,* p. 230.

[35] *Story of the Great American West,* p. 230.

[36] Palmer and Stryker, pp. 61, 73.

[37] *Colombia Encyclopedia,* 4th ed., s.v. "canning," "Appert, Nicholas"; Boorstin, p. 315.

[38] Boorstin, pp. 309, 312-315; Furnas, pp. 690-691.

[39] Bruce Catton. *Grant Takes Command* (Boston: Little, Brown, and Company, 1968, 1969), p. 156.

[40] Furnas, pp. 690-691; Boorstin, p. 315.

[41] Boorstin, pp. 436-437.

[42] Brown, pp. 184, 390-391.

[43] *The World of the American Indian,* pp. 328, 341; Brown, pp. 103, 272, 275-278.

[44] *Story of the American West,* pp. 231, 320.

[45] Palmer and Stryker, pp. 22-23; Henry G. Alsberg (ed.), *The American Guide,* (New York: Hastings House Publishers, 1949), pp. 221-222, 224; Colombia Encyclopedia, 4th ed., s.v. "Hudson River," "West Point."

[46] Palmer and Stryker, pp. 45, 49, 51; Morison, p. 362. "Sappers" were soldiers who laid mines and dug "saps"; the long, narrow trenches used to approach and undermine enemy fortifications.

[47] Palmer and Stryker, p. 51.

[48] Allan Nevins and Henry Steele Commager. *The Pocket History of the United States* (New York: Pocket Books, Inc., 1956), pp. 148-149; Palmer and Stryker, p. 51.

[49] *World Almanac and Book of Facts, 1988* (New York: Pharos Books, Scripps Howard Company, 1987), p. 420.

[50] Manfred Blake Nelson. *A Short History of American Life* (New York: McGraw-Hill Book Company, 1952), p. 308

[51] *World Almanac and Book of Facts,* p. 420.

[52] Blake, p. 308.

[53] Palmer and Stryker, pp. 68-69.

[54] Palmer and Stryker, p. 62.

[55] *Colombia Encyclopedia,* 4th ed., s.v. "Thayer, Sylvanus"; "United States Military Academy"; Palmer and Stryker, p. 64.

[56] William Manchester, *American Caesar: Douglas MacArthur, 1880-1964* (Boston: Little, Brown, and Company, 1978: New York: Dell Publishing Company, 1979), p. 63; *Colombia Encyclopedia,* 4th ed., s.v. "United States Military Academy."

[57] Manchester, p. 62.

[58] Manchester, pp. 62-63.

[59] Manchester, p. 69.

[60] Manchester, p. 69.

[61] Brown, pp. 244-245.

BIBLIOGRAPHY

Government Manuals

General Regulations for the Army of the United States. Washington, D.C.: Government Printing Office, 1821, 1825, 1841, 1857, 1861, 1863.

Quartermaster Support of the Army: A History of the Corps, 1775-1939. Prepared by Erna Risch. Washington, D.C.: Government Printing Office, 1962.

Regulations Concerning Barracks and Quarters for the Army of the United States, 1860. Washington, D.C.: George W. Bowman, Printer, 1861.

Report on Hygiene of the United States Army. Circular No. 8. Prepared by John S. Billings, Assistant Surgeon, U.S. Army. Washington, D.C.: Government Printing Office, 1 May 1875.

Specifications for Means of Transportation, Paulins, Stoves and Ranges and Lamps and Fixtures for Use in the United States Army. Washington, D.C.: Government Printing Office, 1882.

United States Congress. United States Senate. *Medical Statistics of the U.S. Army,* 1839-1854. U.S. War Department.

United States Quartermaster General. *Annual Reports of the Quartermaster General to the Secretary of War,* 1867-1900.

United States Surgeon General's Office. *Report on Barracks and Hospitals.* Circular No. 4. Prepared by John S. Billings, Assistant Surgeon, U.S. Army. Washington, D.C.: Government Printing Office, 5 December 1870.

United States War Department. Corps of Engineers and Department of the Army. *Regulations for the U.S. Military Academy at West Point, New York.* [Various dates throughout the nineteenth century.]

Manuscripts and Intergovernmental Reports

Brown, William L., III. *Pictorial History of Enlisted Men's Barracks of the U.S. Army, 1861-1895.* Harpers Ferry, WV: United States Department of the Interior, National Park Service, 1984.

Brown, William L., III. and Burton K. Kummerow. "A Study of Military Furnishings in the 1820s." Manuscript for Fort Snelling, Minnesota.

Chappel, Gordon. "Barracks Furnishings of the United States Army: The Transitional Years, 1860-1890." San Francisco, CA: United States Department of the Interior, National Park Service, September, 1976.

Clary, David A. *The Relics of Barbarism: A History of Furniture in Barracks and Guardhouses of the United States Army, 1800-1880.* Harpers Ferry, WV: United States Department of the Interior, National Park Service.

Rickey, Don, Jr., and James W. Sheire. *The Cavalry Barracks, Fort Laramie: Furnishings Study.* Prepared for the United States Department of the Interior, National Park Service. Washington D.C.: Government Printing Office, 1969.

Other Publications

Custer, Elizabeth B. *Boots and Saddles.* 1885. Reprint. Norman, OK: University of Oklahoma Press, 1961.

Grange, Robert T., Jr. "Fort Robinson: Outpost on the Plains." n.d. Reprint. *Nebraska History Magazine,* 39, No. 3 (1958): 191-241.

Journal of the Company of Military Historians. Westbrook, CT. Four issues yearly. Published since 1949.

Marcy, Randolph B. *The Prairie Traveler: A Handbook for Overland Expeditions.* New York: n.p., 1859.

Mayhew, Edgar deN., and Minor Myers, Jr. *A Documentary History of American Interiors.* New York: Charles Scribner's Sons, 1980.

Peterson, Harold L. *Americans at Home.* New York: Charles Scribner's Sons, 1971.

—-. *The American Sword,* 1775-1945. Philadelphia, PA: Ray Riling Arms Books Company, 1954.

Prucha, Frances Paul, ed. *Army Life on the Western Frontier.* Norman, OK: University of Oklahoma Press, 1958.

—-. *Military Posts of the United States, 1789-1895.* Madison, WI: The State Historical Society of Wisconsin, 1964.

Rickey, Don, Jr. *Forty Miles a Day on Beans and Hay: The Enlisted Soldier Fighting the Indian Wars.* Norman, OK: University of Oklahoma Press, 1963.

Roe, Frances M.A., ed. *Army Letters from an Officer's Wife.* Lincoln, NE: University of Nebraska Press, 1981.

Seale, William. *Recreating the Historic House Interior.* Nashville, TN: American Association for State and Local History, 1979.

—-. *The Tasteful Interlude: American Interiors through the Camera's Eye, 1860-1917.* 2nd. ed. Nashville, TN: American Association for State and Local History, 1981.

Simpson, Jeffery. *Officers and Gentlemen: Historic West Point in Photographs.* Tarrytown, NY: Sleepy Hollow Press, 1984.

Stallard, Patricia Y. *Glittering Misery.* Fort Collins, CO: The Old Army Press, 1978.

Todd, Frederick P. *American Military Equipage, 1851-1872.* Vol. II, State Forces. n.p.: Chatham Square Press, Inc, 1972.

Winne, Carolyn Frey. "Letters of Carolyn Frey Winne from Sidney Barracks and Fort McPherson, Nebraska, 1874-1878." Edited by Thomas R. Buecher. *Nebraska History Magazine,* 62, No. 1, Spring 1982.

INDEX